REDEEMING
TELEVISION

How TV Changes Christians—
How Christians Can Change TV

Quentin J. Schultze

INTERVARSITY PRESS
DOWNERS GROVE, ILLINOIS 60515

InterVarsity Press is the book-publishing division of InterVarsity Christian Fellowship, a student movement active on campus at hundreds of universities, colleges and schools of nursing in the United States of America, and a member movement of the International Fellowship of Evangelical Students. For information about local and regional activities, write Public Relations Dept., InterVarsity Christian Fellowship, 6400 Schroeder Rd., P.O. Box 7895, Madison, WI 53707-7895.

Cover photograph: Michael Goss

ISBN 0-8308-1383-7

Printed in the United States of America

Library of Congress Cataloging-in-Publication Data
Schultze, Quentin J. (Quentin James), 1952-
    Redeeming television: how TV changes us—how we can change TV/
  by Quentin J. Schultze.
      p.   cm.
    Includes bibliographical references and index.
    ISBN 0-8308-1383-7
    1. Television broadcasting—Moral and ethical aspects.
    2. Television broadcasting—Social aspects—United States.
    3. Television programs—Evaluation.   4. Television viewers.
    I. Title.
    PN1992.6.S285   1992
    302.23'45—dc20   92-12474
                                         CIP

15	14	13	12	11	10	9	8	7	6	5	4	3	2	1
04	03	02	01	00	99	98	97	96	95	94	93	92		

To my students—
past, present and future

Preface

During the past fifteen years I have taught courses on the mass media, especially television, at Drake University, Calvin College, Regent College and Hope College. I've also lectured at many more colleges and universities, and spoken to hundreds of civic, religious and media organizations. These experiences have been an education for me.

One of the things I have learned is that people tend to believe what they want to believe—not what is true or worthy of belief. This is as true for well-educated folk, such as those who have doctoral degrees, as it is for media pundits and, regrettably, people of religious faith. It is quite difficult to get anyone to reconsider what they feel strongly about in important matters such as politics or religion. As a result, the books that sell the best, just like most popular television programs, confirm existing cant, superstition and (sometimes) wisdom.

This book is my attempt to question what a lot of people believe about television. It is also my attempt to provide some answers or at least alternative questions. I hope to challenge viewers and producers, scholars and common souls. I invite all interested parties to join the discussion. This includes pastors, teachers, parents and videophiles of all ages. It also includes people who like to criticize television. In fact, I suspect some of the critics will be most unnerved by what I have to say.

I thank InterVarsity Press for publishing this book. Most religious publishers have focused their efforts on manuscripts that blast the tube with the kind of rhetoric that the Christian community already believes. My editor at InterVarsity, the laconic and good-spirited Rodney Clapp, believed in this alternative project from the beginning. His support and encouragement have been commendable.

At Calvin College I have enjoyed the most remarkable departmental colleagues, who could always laugh at academic pretentiousness masquerading as truth and knowledge. My co-conspirators in the department of communication arts and sciences at Calvin are a rare breed of scholar-teachers. They are not just amiable, but downright fun to work with, even when there is too much to do and too little time. I'm delighted to report that these modern-day Calvinists are a source of grace and good cheer—even when the TV set is on in the office.

My dear friend Lois Curley gave me much counsel on this book. She has a gift for seeing the underlying gist of a manuscript and communicating it to publishers. I'm thankful for her.

Readers will detect that this book provides examples and illustrations from my family life. My wife, Barbara, and two children, Stephen and Bethany, have taught me a lot about television and the family. They deserve public commendation for instilling confidence in their father and husband—even when they told him how wrong he was.

Finally, I have dedicated this book to my students. They have undoubtedly taught me more than I have taught any one of them.

But to the Creator, Redeemer and Comforter I present any and all glory. Grace is a joyful mystery.

Introduction

I grew up in the first generation reared on television. My father brought home a large Philco set in a mahogany cabinet. It rested rather unsteadily on a black cast-iron stand with long, thin legs. As I recall, our floor-model radio was moved to the basement to make room for the new tube.

The TV quickly became the focus of evening leisure. We rearranged the furniture for viewing, transforming the parlor into a home theater.

There was no "Sesame Street" in those days, but I watched plenty of other shows: "Dobie Gillis," "Garfield Goose," "Captain Kangaroo," Bishop Fulton J. Sheen, "Leave It to Beaver," "The Honeymooners," "Red Skelton" and many more. Late at night, while I lay in bed listening to sounds drifting from the parlor, I tried to figure out why my father was laughing. Years later I found out he'd been watching Ernie Kovacs, a brilliant, iconoclastic comedian who pioneered low-tech special effects.

Soon I figured out that not everyone shared my enthusiasm and my parents' lenient attitude about viewing TV. Some neighbors did not appreciate television's "worldly" programs. My playmates were not always allowed to watch the shows I liked. A few parents even refused to buy a set. They saw the tube as an electronic Trojan horse.

Several decades later, TV has not changed as much as most people believe. Parents are still divided over what shows to let their children watch There remains an amazing diversity of programs, including the return of some of my old favorites. No one could have predicted the amazing popularity of "Sesame Street," the growth of cable and the VCR, and the business of televised sports, especially football. But I had "Romper Room," the first portable sets, UHF channels, color TV and even professional wrestling shows. My father watched

Hulk Hogan's televisual ancestors. So did I, in utter childish confusion.

This book is written for everyone, like me, who enjoys television but would like to make it better. The book is not a wild-eyed attack on sex, violence and profanity. Instead, it contains realistic suggestions about how to redeem the tube. My goal is to help readers become active viewers, not just passive watchers, of the tube. Along the way, the book offers specific suggestions for parents, teachers and college students, as well as for people who work in television. We need more committed Christians in all phases of the business, from public television to the newer cable networks.

In America today there is a lot of heat surrounding television, but little insight. One day television is the one-eyed monster or the boob tube—what former Federal Communications Commissioner Newton Minnow called a "vast wasteland." The next day television is a source of national pride or a vehicle for celebration, broadcasting the Super Bowl, the inauguration of a president, or humankind's first steps on the moon.

Consider what happened on American television from 1987 to 1991. The tube captured the quivering lips of fallen televangelist Jimmy Swaggart as he appealed for forgiveness to his Baton Rouge congregation and supporters around the globe. In Philadelphia, a local station aired a graphic tape of a city official as he put a pistol in his mouth and pulled the trigger. "60 Minutes," one of the highest-rated programs, temporarily destroyed the market for apples with an exaggerated report on the potential hazards of a pesticide some growers used. Cable News Network (CNN) provided millions of viewers live coverage of the public protests against the Chinese government at Tiananmen Square. Public television broadcast a remarkable Civil War documentary that riveted millions of Americans to their sets and elicited a barrage of discussion about the meaning of the war in the nation's history. And CNN's dramatic live reports captured the first barrage of American and allied bombs on the city of Baghdad—an amazing display of nighttime fireworks that launched the short-lived war in the Persian Gulf.

For every criticism of television there are thousands of exceptions. This book challenges the critics *and* supporters of television. Both tend to emphasize only one part of the story. Television, like the rest of creation, echoes the goodness of God's grace daily. Unfortunately, at the same time it reflects the dark shadows of human selfishness, rebellion and pride. Television is a potentially great medium that suffers in the hands of sinful users and misinformed critics.

People who tout the benefits of television generally are blind to the selfishness of some television industries and the sinfulness of viewers. In short, television's most enthusiastic fans, including many televangelists and their supporters, seem to have a very inadequate view of the Fall; they seem to believe that technology is the key to accomplishing grand purposes such as evangelizing the world and educating people's minds.

At the same time, television's critics, evangelicals included, are too fearful of the medium's apparent wickedness. Popular critics like Malcolm Muggeridge, Neil Postman and Jacques Ellul fall into this category. In their eyes, television is almost the devil incarnate, a nefarious conduit of triviality and deception. If television's biggest supporters overlook the devastation of the Fall, its critics miss the goodness of God's creation.

It's time to remove the self-imposed blinders that Christians use to make sense—and nonsense—of the tube. We need an understanding of television that will help us use the medium wisely without falling victim to either the triumphalism or the despair that too often characterizes Christian use of this important technology.

My thesis is that television can be "redeemed" when producers and viewers alike hold the medium up to standards of spiritual, moral and artistic integrity. I argue that we do not simply need more televangelism or even necessarily more "Christian TV"—if by such a term one means only programs that proclaim the gospel or are produced by evangelicals. Rather, we need a much clearer sense of the many ways that television can glorify God and serve humankind.

Chapter one reveals the implicitly religious ways that people think about television, especially in North America. On one side are the TV preachers and secular champions of the tube who seem to have more faith in the technology than in God. On the other side are the many critics, like Postman and Muggeridge, who appear to have lost their faith in God's ability to use television in ways that would truly serve humankind. These opposing groups fail to see the tube in the context of *both* the creation and the Fall. They forget that television is not only a technology, but also a social institution with its own values and professional practices. Discerning viewers and conscientious producers both need to know these basic distinctions.

Chapter two surveys the breadth of possibilities for using television to tell stories. Nothing has kept Christians and others from redeeming television more than the misguided belief that every program must teach or instruct people about the gospel of Jesus Christ. In evangelical circles, the narrow linking of the tube with the Great Commission has resulted ironically in few programs that actually reach nonbelievers. Television, as a narrative medium, can be used effectively to amuse, instruct, confirm and illuminate. Whatever stories can do, television can also do—and surely it is the most popular storyteller of our age. If television is truly going to be used to spread the good news, Christian producers must be creative storytellers through drama and documentary, not just Bible teachers or talk-show hosts. In short, believers must seize the medium for the breadth of its cultural potential, not just for its apparent power to propagandize.

As chapter three shows, North Americans have learned not to expect much from television. Amusement is the dominant role of the medium in most peoples' lives, and viewers are generally happy to get a few chuckles from the one-eyed jester. We *watch* TV instead of *viewing* it. And the commercial television industry encourages this by making shows that often aren't worth viewing. Our expectations—our very sense of the medium's potential—have been badly stifled for most viewers by the traditional ways the tube has been used.

Chapter four is a primer on how to view television. Because the medium is different from the stage or even film, viewers need to know the tube's visual language—how it communicates. An understanding of the medium goes a long way toward reducing its negative effects on us. Viewers ought to be just as televisually savvy as program producers. Although we live in a visually oriented culture, most of us do not understand how moving-image media communicate. Historically speaking, Protestants were especially uneasy about potentially idolatrous images and icons, a fear that still colors some Christians' feelings about all visual communication. While such concerns are understandable, they are also lamentable, for they have greatly impoverished Protestant attempts to redeem television. Today, explicitly nonreligious television programs fill the need among Americans for such images and icons. The tube's visual language—its *iconography*—is naturally biased toward personas, from the newscaster to the series star to the religious celebrity. Chapter five offers standards for evaluating television drama. It summarizes social-scientific theories on the effects of particular types of programs. The chapter also discusses the benefits and liabilities of using television reviews in secular periodicals. Finally, the chapter addresses one of the most difficult issues of quality: the redeeming value of programs that are not explicitly Christian.

The sixth chapter looks at how Christians wrongly equate immorality only with sex, violence and profanity. While claiming to be champions of morality, believers often ignore racism, ethnocentrism and materialism. Also, Christians forget that the message of a show is not the same as its simple depiction of events. A show can condemn sexual promiscuity as well as condone it; the scenes alone are only part of the story. Christian viewers must be much more informed about how the tube communicates morality. Otherwise they will continue to be stereotyped as moralistic prudes who do not enjoy life as created by God. And Christians will be exploited by moralistic watchdog groups that simplify issues in order to raise funds.

Chapter seven examines critically the people and organizations that make most programming. After all, television's redemptive potential always is limited by the minds and hearts of the people entrusted with the medium. Programming is shaped by the allegiances as well as the abilities of writers, directors and especially producers. We cannot select programs and view them wisely if we are ignorant about the people and organizations that make them.

In commercial television, the producers are hardly free to exercise their own wills. The demand of the viewers, reflected quantitatively in audience ratings, shapes the content of all programming. As a result, much television drama is made by people with "holes in their souls," to use Julia Phillips's phrase— people who are willing to make whatever shows will sell. Advertiser boycotts and network letter-writing campaigns can sometimes influence producers and broadcasters, but not as much as audience ratings. If Christians do not understand this system, they cannot effectively redeem it.

The last chapter returns to the broad picture. It calls Christians to claim television discerningly for the glory of God and to the benefit of humankind. This means that we must inspire talented people to enter the "secular" television industry, establish explicitly Christian media, encourage each other to view the tube discerningly, and cultivate in our communities knowledgeable Christian critics who can provide wise evaluations of programming.

Television is worth redeeming. It is, after all, a gift from God discovered by human beings. Once we get past the naive praise heaped upon the medium by its uncritical supporters, and the cynical condemnations nailed to television by its elitist detractors, there is indeed room for more redemptive television. The electronic Trojan horse can be transformed into a chariot of grace.

1
THE GREAT COMMOTION/

Why the Christian Critics & Advocates of TV
Are Usually Wrong

*The LORD God took the man and put him in the Garden of Eden to work it and take
care of it. (GEN 2:15)*

A NUMBER OF YEARS AGO I FIRST READ MALCOLM MUGGERIDGE'S *CHRIST
and the Media,*[1] a provocative indictment of television. According to the
former BBC-TV personality, the tube inherently distorts reality and tells lies.
Although Muggeridge's argument was persuasive, it troubled me. Both as a
Christian and as a communications scholar, I felt that there must be something good about television. Surely not all programs were evil.

Nevertheless, Muggeridge's point was not easily dismissed. The airways and
cable channels are filled with programs that seemingly promote anti-Christian
values and beliefs.

So I was shocked several years later to see that Muggeridge was going to
appear on William F. Buckley, Jr.'s "Firing Line" show. Why would one of
the tube's outspoken critics agree to go on the air? Did his appearance indicate
a change of heart about television? I tuned in with great anticipation.

For two Sunday afternoons in a row, Muggeridge gave an eloquent testi-
mony of his faith in Christ. According to typical television standards, the

programs had almost nothing to offer—no action or adventure, no sex or violence, no fancy graphics or rapid-fire editing; simply two adult males sitting in chairs and discussing matters of life and faith. Yet from my perspective, the programs powerfully communicated the authenticity of Muggeridge's character and the relevance of Christ for the human condition in the contemporary world.

Muggeridge's appearance on television helped convince me that his blanket condemnation of the medium was badly misguided. The next step was to figure out how he had erred in his assessment of the tube. I found that Muggeridge, like many non-Christian critics of television, greatly oversimplified matters: He had part of the truth and tried to make it the whole truth. In many respects Muggeridge's condemnation of television parallels Neil Postman's thesis in his popular book *Amusing Ourselves to Death*.[2]

Eventually I realized that Christians' views of the tube often have been shaped by the surrounding culture, not by distinctly biblical perspectives. Strangely enough, this is the case even for many advocates of so-called Christian TV, not just for the critics. Surely television deserves substantial criticism. As Ashley Montagu once wrote, the tube has become "the god of the common man's idolatry, his oracle, and the principal source of his news and entertainment."[3] This kind of power, even if overstated, requires an enormous responsibility. The television moguls have not always taken this responsibility seriously. Television should not "hold up a mirror for man to see himself merely as he is and to maintain himself in that image," says Montagu. "On the contrary, it should be the function of television to exhibit the image of man as he can and should be."[4]

Because we have not held television up to such a standard, there is much more commotion than clarity about the medium and its legitimate role in society and church. It's time for a fresh look at the most popular mass medium in the world today. And it's time to do it with biblical glasses.

The Burden of the Great Commission
Some of the Christian commotion about television stems from American evangelicals' theological emphasis on Jesus Christ's Great Commission to his followers: "Therefore go and make disciples of all nations, baptizing them in the name of the Father and of the Son and of the Holy Spirit" (Mt 28:19). Although evangelicals have rightly kept alive the fundamental necessity of

personal religious conversion, they have also sometimes tried to make such conversion the alpha and omega of *all* Christian activity in the world. Too often "Christian TV" is equated with missionary television.

American television has suffered considerably because of this linkage of "Christian" and "evangelistic." By using these terms interchangeably, and by assuming that the primary purpose of television programming should be to convert viewers to Christ openly, American evangelicals ironically have created their own broadcast ghettos that rarely attract nonbelievers. The vast majority of contemporary televangelism was never an evangelistic triumph.[5] In fact, some of the most popular programs, such as "The 700 Club," became less preachy and more feature- or news-oriented over the years.

The irony is clear: Programs that are less evangelistic generally attract more viewers. CBN's satellite-cable network, eventually called "The Family Channel," rose in the industry when it replaced much of its preachy programming with old situation comedies, Westerns and family-oriented Hollywood movies.

An overemphasis on the Great Commission, then, has sometimes become a burdensome albatross around the necks of producers and viewers of Christian TV. It has tended to turn Christian TV into commercials for Christ, veritable sales pitches. As Kenneth Myers puts it, "With the overarching demand of the Great Commission looming over every evangelical enterprise, it is very easy to go for ratings instead of rationality, quantity rather than quality."[6]

This emphasis has also contributed to evangelical confusion about the purpose of art and entertainment in human life. Hans Rookmaaker once wrote that "art needs no justification."[7] Gene Veith, Jr., has said that the "arts . . . are built into creation, into the divinely fruitful fabric of existence and into the human being as the image of God."[8] How many evangelicals believe instead that evangelism is the only real justification for popular art, including television?

I have witnessed how the preoccupation with the Great Commission has led to misplaced and often unfair criticism of talented Christians who sought to improve "secular" broadcasting. A sixty-year-old evangelical friend, highly successful in broadcasting and convinced that he has never compromised his beliefs, tells me that his mother still cannot understand why he would have entered such an evil business. An influential producer of network fare tells a similar story of frustration with other Christians who assume that he must have sold his soul to the devil in order to get ahead in the business—unless he is there to save the souls of his coworkers. Actually, some of his programs deal openly with biblical themes in prime time on the major networks. Why should he have to defend himself for failing to preach the gospel on television? For failing to convert his colleagues? Did he not have a broader calling than evangelization?

Narrow-minded appeals to immediate evangelization have undoubtedly stifled creativity, resulting in artistically inferior programming. Christian TV is among the most predictable and hackneyed programming on the tube. Often in the name of evangelism, Christian video has turned the power of the gospel into an unbelievable tale. The last thing we need is more idealized conversion stories or docutestimonials that distort the way the Holy Spirit normally works in the lives of real people.

I believe that part of the power of the Muggeridge broadcast was its simple explication of the unadorned life of faith. It communicated the gospel in the context of one person's authentic spiritual journey, not in some vain, show-business performance slickly designed to capture souls for Christ.

This is not to say that television should never be used by the churches for evangelism or education. Nor is it to say that evangelism is unimportant or even secondary in the mission of the church. When we begin with the Great Commission, however, we wrongly assume that the only viable and worthy way to redeem television is with more preaching and teaching. Actually, _everything_ we do in communication is implicitly evangelistic as a witness to the lordship of Jesus Christ. The ways we conduct business, handle interpersonal relationships and perform drama are implicit statements of faith. The Great Commission can be fulfilled only with the recognition that all culture is a witness to the beliefs of those who create it. Because of the way God created humankind, all of life is evangelistic.

The Cultural Mandate (Gen 1:28) is the context for the Great Commission. God's claim is not over just human souls, but over the entire creation—including, as John Calvin put it, "all the arts," which "come from God and are to be respected as Divine intentions."[9]

Suppose we evaluated all of our communication—everything we write and say—solely in terms of whether or not the gospel was directly proclaimed. Most college lectures would be unacceptable. So would most musical recordings and concerts, mathematics and engineering. In fact, if we said "I love you" to someone it could be construed as non-Christian.

When the Great Commission is our alpha and omega in communication, we fall into naive views of television. Actually, everything we do has an evangelistic dimension in the sense that all our actions reflect our "witness" (or the lack thereof) to the lordship of Christ. But it is simple-minded to evaluate all communication on the basis of whether it was intended to directly proclaim the gospel or even to teach the Scriptures.

Communication as Vocation

Christ's redemption extends far beyond saving our souls. We must see the Great Commission and all other human communication in the context of God's creation. Humankind's primary vocation is to obey God, to be responsible servants in all aspects of our lives, not only to serve the specialized callings of evangelist or pastor.

Too often Christian TV is reduced to *missionary* television. When this happens, television producers are expected to be evangelists, and viewers are seen as sawdust-trail audiences. This is precisely why most evangelicals who go into television are preachers and teachers.

Since Adam named the animals, humankind has been using communication to do far more than spread the gospel. The Cultural Mandate given to humankind in the first few chapters of Genesis established *all* people's vocation in life: to tend the creation and develop it responsibly. This is indeed what we all do, for good or bad, and for work as well as play. In short, God created humankind to be caretakers and developers of his world—to have dominion over the plants and animals and everything else.[10]

The Latin root for the word "communication" *(communicare)* means "to share," or literally "to make common." It has the same root as "communism" (having property "in common") and "communion" (having the sacrifice of Christ "in common" as we celebrate the sacrament). Christianly speaking, communication is first of all the process of creating and sharing a Christian culture on earth. We do this by using symbols, including words and images, which convey meaning.

Communication enables us to fulfill the Cultural Mandate by working and playing *together* as God's image-bearers. It's hard to think of a single human task that does not require the ability to "share culture"—to have common "ways of life," in T. S. Eliot's phrase.[11] As Raymond Williams once put it, communication is not secondary in human affairs, but necessarily primary.[12] Society depends directly on communication for its very existence. Without communication, our lives would be impoverished and we could accomplish little. There would be no architecture, schooling, art, business and mass media. By God's design, all "common" human endeavors are made possible by human communication. Ironically, the church's tendency to restrict its notion of communication to evangelism actually parallels important shifts in secular Western society. In the last few centuries, Western conceptions of communication shifted gradually from the idea of "sharing" to the concept of "controlling."[13] This led eventually to the popular view that communication is mere transmission and reception—the "sending and receiving of messages."

By the time of the telegraph in the late nineteenth century, few people still viewed communication as a means for establishing a common culture. Communication and travel were nearly synonymous.[14] Growing numbers of people perceived communication as a means for sending economic and political information faster and more profitably across geographic space than ever before. The telegraph would give a competitive edge in the stock market or futures trading; it was a way of achieving personal or corporate gain, not so much a means of contributing to the nation's common culture.[15]

As a result, the mass media were increasingly identified only with manipulation and control—getting other people to think or do things. Modern industry was partly responsible for this shift. Contemporary notions of communication are greatly influenced by the practices of advertising and public relations, which have made "communication" and "persuasion" nearly synonymous. The church increasingly imitated the communication theories of modern business, especially after World War 2. Evangelicals, in particular, quickly adopted the "manipulation" mode of communication because it seemed to fit so well with the emphasis on the Great Commission. They used secular models of communication and marketing to reach the world for Christ.[16] Along the way, they lost sight of the scope of communication in the church's broader task of fulfilling the Cultural Mandate. The church largely turned the culture-forming process over to other social institutions, including the media.

The fact is that communication is central to practically everything we do—work, play, worship—and not only to explicitly persuasive activities. All areas of life depend on the use of communication to maintain culture. Family meals, schoolroom lessons and bedtime prayers are (or should be) meaningful rituals of communication.

Communication enables us to share our lives with others in both profound and mundane ways—to worship with others, to work with others, to get to know others, to break out of our loneliness and alienation with each other and God. What would be more destructive to our godly vocation in life than to deny us our ability to communicate? How little we would accomplish, and how meaningless life would be for us!

In the context of our responsibility to God, television can help us to enjoy, take care of and develop God's creation. Any communication that furthers God's interests in this world is Christian.

Though television is "good" in God's creation, it is never an end in itself. The technology does not exist for its own sake, but for the kingdom of God. Later I will look more closely at what it means to glorify God in communication that is not directly or specifically evangelistic. For now, let me simply suggest that Christian communication is necessarily responsible, and to be responsible is to establish and maintain the kind of culture that glorifies God. Humankind's vocation is not only to evangelize but also to be good and faithful stewards of the creation. As Merrill R. Abbey once wrote, Christians should be "faithful stewards of the media."[17]

All of our communication and culture are witnesses to the gospel—at least they *should* be. "Christian television" includes far more than televangelists, gospel music programs and conversion dramas. It means much more than persuading people to accept Christ as their Lord and Savior—though it certainly includes that.

Christian Critics

Another reason Christians have not used television to its potential is that many Christian leaders apparently believe that the technology is inherently evil. Among the more insightful and cogent critics are Muggeridge, Virginia Stem Owens (author of *The Total Image)* and Jacques Ellul, whose book *The Humiliation of the Word* is a blistering attack on all image-making from the fourteenth century to the present.[18] Owens artfully attacked the church for mindlessly adopting the techniques and methods of modern marketing and promotion, especially as they are used in the television industry.

These critics' work is the Christian version of fears about television represented more broadly by Jerry Mander in *Four Arguments for the Elimination of Television*[19] and by Neil Postman, especially in *Amusing Ourselves to Death: Public Discourse in the Age of Show Business*. These are some of the most trenchant and popular prophets of television doom, and their ideas have influenced the attitudes of many Christians, especially evangelical educators.

These critics share the belief that television might be unredeemably evil because in the process of being used it promotes deception. By themselves the lights, cameras and editors can do little harm; once the camera is turned on and the audience is in place, however, evil is set loose. For them television is not like a gun, whose wickedness apparently depends upon how it is used: to defend an innocent victim against a homicidal attack, to fight a communist war of aggression, to take target practice in the backyard or to hunt deer on government land without a permit. In the view of Owens and Ellul, television always promotes evil—even in the hands of Christians—*because the tube necessarily communicates lies*. As Ellul puts it, the consequences for the church are devastating when believers put "religion in this [false] reality."[20]

Muggeridge says that a television camera cannot easily communicate truth. Unlike written words, which must be composed by someone in particular, the televised or filmic image "is machine made. . . . it is seeing with not through, the eye; looking but not seeing. . . . It's very nearly impossible to tell the truth in television. . . . If you set up a camera and take a film, that is not considered to be anybody's views."[21]

In other words, the camera lies because it must pretend to be somebody's message when in fact it is not any single person's ideas or thoughts. Yet, from the viewer's perspective, television is showing "real" images regardless of how fabricated or fictionalized the content of the program. Viewers may not believe all they see or hear on the tube, but television provides no human context—no "author."

Television creates the impression that the images are not *from* someone's mind or heart, but that they are simply reproductions of reality. The result, says Muggeridge, is that television "has enormously interfered with . . . communication between men, distorted it, deflected it, making it, on the one hand, from the Devil's point of view, advantageous in that it facilitates deception, and on the other, making it the more difficult to communicate those things that are true and real." As Owens puts it, television cannot "catch reality"; the tube is capable of communicating only "directly apprehended, unmediated experience."[22]

These influential critics believe that televised communication is fundamentally different from the written or spoken word. To the viewer, televised messages appear lucid, direct and understandable. As people used to say, "What you see is what you get." Therefore, television is simply *experienced* by viewers, who do not think about who made the programs, why they were produced and whether or not they are true.

If someone greets us on the street, we consider the meaning of the greeting in the context of where we are, who issued the greeting, exactly what was said, the time of day and sometimes even the weather. When reading a book we frequently stop to think about what we are reading, who wrote it, the feelings generated in us by the book and how its message relates to our own experiences. Rarely do we engage television in the same ways, say the critics of the medium.

Although their arguments are frequently overstated, television critics point to an important aspect of the medium: its amazing ability to create the illusion that there is no one "behind" the tube and, as a result, to make a program's content appear to be reality. When someone is married on a soap opera, hundreds and even thousands of viewers will send cards, letters and gifts to the newlywed characters.[23] For years Walter Cronkite was deemed the most trustworthy man in the United States by the American people, few of whom had ever met him. After the end of a long and successful run on network television, "M*A*S*H" left the air amid the tears and memories of millions of faithful viewers; as many people in the final audience said, the characters on the program seemed like friends. Similarly, millions of viewers tune in their favorite televangelists, whose faith seems to be especially genuine and inspiring.[24]

These critics are partly correct. Television's visual appeal can deceive viewers. It is not easy for most viewers to think about a program while they are being bombarded with images. By the time the program is over, the images are gone—unless one has taped them. Most of the images and dialog are only vague memories that can be recalled with little clarity and precision. As a result, even in educational settings it is very difficult to get audiences to reflect on and talk about the images of a television program or film.

This makes the tube a highly manipulative medium. In its programming, the tube seems little more than an entertaining or informative source of pleasure and delight. In fact, though, television is a medium of communication with its own visual logic deployed by an array of writers, directors and producers.

Moreover, the critics are right that television rarely offers an identifiable individual's point of view. Like cinema, television is a collaborative medium that requires the contributions of many people for the making of even the simplest programs. It's not easy for a viewer to become more critical and thoughtful about who is actually communicating on television, precisely who is responsible for the programming and exactly whose views are being communicated.

Even with the televangelists, the viewer can't be sure that the preacher believes his own message; scripts are sometimes written to elicit audience emotions and generate contributions, not simply to convey a biblical message. Meanwhile, the tube creates the impression that the preacher is talking personally to the viewer.

The Gnostic Impulse: TV as Inherently Evil

But the critics err when they assume that television in itself is a deceptive medium. Every medium, from speech to radio to film, offers problems similar to those identified by Muggeridge, Owens and Ellul.

A corporate official may deliver a speech written by her public-relations staff but presented as the official's personal views. The campaign staff of a presidential candidate may carefully select the location, setting, props and music for a campaign appearance. Newspapers and even novels can be highly propagandistic and thoroughly deceptive about historical events or contemporary issues.

At best, we can say that the written word, because of its potential for precision and because it is recorded on the page, is more open to critical reflection than the spoken word or transient images of screen and tube. The fact that most television viewers are unreflective, however, may have more to do with the kinds of programming available and the lack of critical visual education than with the inherent nature of television.

The argument that television is inherently deceptive usually dissolves into the premise that the technology cannot communicate real spiritual truth, especially the gospel itself. Owens writes boldly that "the camera cannot hope to catch" that with which "the gospel is concerned."[25] In her view, the gospel is too "grotesque," its claims too "extravagant and incongruous," for television's simplistic messages and shallow personalities. She argues that any attempt to shape the gospel to fit within the limits of the medium will simplify and distort the message; the best television can do is to convey information about the church or about particular personalities within the church.

For Muggeridge, the media have power "only to the extent that they can influence and exploit the weaknesses and wretchedness of men—their carnality which makes them vulnerable to the pornographer, their greed and vanity which delivers them into the hands of the advertiser, their credulity which makes them so susceptible to the fraudulent prospect uses of ideologues and politicians; above all, their arrogance, which induces them to fall so readily for any agitator or agitation, revolutionary or counter-revolutionary, which brings to their nostrils the acrid scent of power."[26] In Ellul's terms, *"Visible reality transferred to the illusion of images becomes our ultimate reference point for living."*[27]

In spite of the contributions such critics make to our understanding of the nature of television, their passionately overstated arguments must be placed in historical context. First, there is a long history among intellectuals of fear and distrust of industrial and technological change.[28] In Western society, intellectuals have frequently bemoaned the eclipse of past ways of life with the coming of new technologies. Believing that something long-cherished was being extirpated by new machines or media, they appealed romantically to the past as a way of protecting society from the potential ravages of the future.

Mass printing was criticized by the Latin-based ecclesiastical establishment of the 1400s, for the religious authorities realized that vernacular publishing could destroy traditional forms of church organization and control. Similarly, in the early 1800s the professional clergy, trained in Eastern seminaries and central figures in the literary establishment, criticized the new breed of itinerant evangelists who used popular music and emotional sermons to draw large audiences.[29] In both cases, defenders of the faith appealed to traditions of the past and to the media with which they were most comfortable to ward off the attack from the new media. In both cases they lost.

Second, intellectuals within the church are often highly suspicious of visual communication. Steeped in the printed word, and often members of the literary or at least the academic establishment, they neither understand nor accept the iconography (visual language) of image-based media. They are comfortable with printed words and deeply suspicious of images, especially mass-consumed images.

This distrust of visual communication can be traced directly to the Reformation, which, for all of its theological and ecclesiastical strength, overreacted to the use of icons in faith and worship.[30] As the new Protestant religious establishment developed in the 1600s, it increasingly moved toward a culture of the printed word and away from an oral or visual culture. Eventually sermons were composed first in writing and then delivered to churches orally. The new Protestant church buildings were increasingly austere, and the interiors were organized for lectures. Worship became synonymous with listening to a carefully crafted, literary sermon. Partly because of the ascetic influence of Calvinism, Sunday clothing was kept simple and plain, and Lord's Day activities centered on personal or communal reading.

The sentiments of television critics within the contemporary church are redolent of the historic ties of Protestantism to a literary establishment. Because visual symbols appear less semantically precise than words, more open to emotional manipulation and generally more popular across social classes, critics of new visual media have often disdained icons of any kind. The results of this hostility toward visual communication are clear in the late twentieth century: visually impoverished worship services, little progress in contemporary Christian visual arts and naive, often misguided criticism of the secular visual arts and media.

Too often the critics of television are overreacting to the obvious mendacity of so much television programming: the arrogance of news reporting, the ungodliness of prime-time drama, the materialism of commercials. There is much spiritual fraudulence in such programming. However, a lack of worthwhile programs should not lead us to condemn the technology per se.

Sometimes the critics sound like British Luddites, who hoped in the early 1800s to protect their textile jobs by smashing new labor-saving technologies. At their worst, Christian arguments about the evil nature of television technology lead to a modern gnosticism. If the critics from the church's intellectual establishments are not careful, their powerful condemnations of television could become one more gnostic heresy against some of the riches of the creation. The difficulty is to maintain an informed, critical approach to all media while joyfully determining how best to use every medium for the glory of God.

The Idolatrous Impulse: TV as Redeemer

If the critics of television too easily condemn it as inherently evil, the Christian advocates of television often represent the opposite end of the spectrum: Television is providentially redemptive. The critics are connected historically with technological pessimism in America; they fear that the medium is destroying the moral and spiritual fabric of society. The technological optimists, represented especially by the televangelists, invest in the medium their deepest hopes and sincerest desires: that the church will seize control of the latest technologies for the glory of God. They believe that God providentially provided the church with television in order to broadcast the gospel around the globe just before the Second Coming of Christ.

There is a long history of American celebration of communications technology—what James W. Carey and John J. Quirk have called the "mythos of the communications revolution."[31] In the United States, people have joyously admired new technologies at fairs and expositions and have enshrined them in museums. Optimistic America associated its rapid industrial expansion and technological development with the will of God.

Not surprisingly, evangelical Christians were leaders in adapting new technologies to mass communication, beginning with book publishing in the early colonies, Bible and tract printing and distribution in the early nineteenth century, periodical publishing during the rest of that century, radio in the 1920s and 1930s, television in the 1950s and satellites in the 1970s and 1980s.[32]

The idea that a new communications technology is providentially redemptive is hardly novel; at least some evangelicals have believed this about every new medium that came along. American Christians always viewed the latest communications media partly in terms of their potential power to transmit the gospel across political and cultural boundaries.

Jimmy Swaggart, one of the best-known Protestant televangelists with a daily or weekly program, was an outspoken critic of televised sex and violence. But he was also one of the major supporters of the church's use of TV technology. "For the first time in history God has given a handful of men the opportunity to reach tens of millions with the gospel of Christ," wrote Swaggart.[33] Of course he saw himself as one of those men, and his optimism about the technology cannot be separated entirely from his own ministry's goals. But all evidence suggests that Swaggart genuinely believed that television could "redirect a nation to the paths of righteousness" when it was in Christian hands.

At one point, before his involvement in a sex scandal in 1987, Swaggart told his viewers that his own broadcast ministry might be able to broadcast the gospel to every nation on earth prior to the return of Christ. It was as if the Second Coming of Christ depended upon Swaggart's own programming.[34]

The most telling articulation of the hope for television's providentially redemptive role in history came from Ben Armstrong, former executive director of the National Religious Broadcasters. In his book *The Electric Church,* Armstrong argued that the newest communications technologies, including satellite and cable television, would become "a revolutionary new form of the worshiping, witnessing church that existed twenty centuries ago."

Calling the technology of broadcasting one of the "major miracles of modern times," Armstrong envisioned a new church—supposedly more like the early church—where members lived in peace and harmony. "Radio and television have broken through the wall of tradition we have built up around the church," he wrote, and "have restored conditions remarkably similar to the early church."

Armstrong wondered, too, whether the angel referred to in the book of Revelation (14:6) might actually be a communications satellite used by God to fulfill the prophecy of the last days: "And I saw another angel fly in the midst of heaven, having the everlasting gospel to preach unto them that dwell in the earth, and to every nation, and kindred, and tongue, and people" (KJV).[35]

If the critics like Muggeridge romanticize literary culture, the television advocates generally romanticize their own populist culture. The tube symbolizes opposing cultural ideals, even if the technology could not really produce such cultures itself. As Willard Rowland, Jr., has cogently argued, the rise of television generated "extreme polls of evaluation . . . both messianic and demonic." American progressives saw TV as "the repository of hope for a revived democratic process, a stronger set of social bonds, a richer cultural life, and a vastly improved educational system." Meanwhile, "established institutions and brokers of morality and values" viewed television as the "latest and most dangerous in a series of technological and social inroads on their authority and status."[36]

In Christian circles, this battle pits the educated clergy and Christian academicians against the Christian communications industry and popular evangelists. There are many exceptions and a host of variations on the theme, but the pattern is clear: A person's standing in the church and surrounding culture generally predicts how he or she will view the technology of television.

Moreover, both sides of the debate have their own collaborators in the surrounding culture. The critics have dozens of supporters and critics in seminaries and universities as well as publishing houses and denominational offices. Meanwhile, the advocates can call on the vast plethora of popular articles and anecdotes about the benefits of every new communications technology: TV creates a new "global village"; the VCR liberates the individual viewer from the tyranny of network programming and scheduling; cable television gives the viewer thousands of real program choices; programming has never been better; society is more moral or more advanced.[37]

If the critics tend toward gnosticism, the advocates gravitate toward idolatry, or at least the veneration of technology. Both positions are strongly entrenched in American culture, so it's unlikely that the church will be able to free itself completely of the burden of each. But viewers must see beyond both myths if they are to redeem their own use of television.

Television as Technology and Institution
Although they come to nearly opposite conclusions, the critics and the celebrators of television share a mistake: They both fail to distinguish between the *technology* and the *social institution* of television.

Every communications medium is first of all a technology. In the case of television, the technology makes moving images (live or recorded) and accompanying sound in one time and place for "recreation" at a different time and place. All the elaborate equipment, from cameras and lights to switchers, editors and computers, is used for that rather simple technical goal. The technology of television, then, is merely the equipment used to create, store and transfer the moving images and sounds.[38]

As a result, the technology is inherently neither good nor bad, neither something to idolize nor something to condemn. It is mere equipment whose potential is limited by the finitude and depravity of humankind as well as the technology itself.[39]

The potentially redemptive qualities of television are clear only when the technology is used by people. Television without action is benign, but it is also not completely part of the riches of the kingdom of God. Our vocational responsibility is both to develop this technology and to use it wisely as servants of the Creator. Only then does television have any real value for God and for humankind.

Once television is used, the technology quickly becomes part of one or more *social institutions*. The institutions of television include all the values, beliefs and actions that guide our use of the technology. Embedded in society, technology becomes a "distinct human cultural activity in which human beings exercise freedom and responsibility in response to God by forming and transforming the natural creation."[40]

In other words, social institutions establish *why* people use television as well as *how* they use it. For example, in the United States the commercial television networks, local stations, production companies and advertising agencies are important organizations. They have decided that the technology of television should be used primarily to attract audiences and to sell those audiences to advertisers. In one way or another, nearly everything on commercial TV that is done with cameras, editors and microphones reflects the goal of building audiences for advertisers.

By contrast, in the United States, Canada and Great Britain, public television has been used primarily to educate and inform audiences—however small those audiences might be. This is why, for instance, many American public stations air programs about how to fix up old houses, understand architecture and appreciate the arts.

Commercial and public television use the same technologies, but because of the different values, beliefs and practices of the people in charge, the content of their programming is quite different. And so is the effect of their programming on society and the individual.

But there is even more to television's social institutions. Families have established patterns of viewing—when, where, why and how much to view—that play a very important role in the institution of modern television. Families have implicit values that influence not only what they watch but also what commercial networks are likely to air. Audience ratings are part of the institution.

Finally, the institution of television nearly always has its own officials who make judgments about what is proper programming. In most societies this role is held by the government or a quasi-governmental agency, such as the Federal Communications Commission in the United States. But it is also a role played by television critics who write for newspapers and magazines or voice their program evaluations on radio and television.

The *medium* of television is always a combination of technology and social institution. The technology determines the ultimate potential of television—what could possibly be communicated—while the institutions establish the practical limits (for example, what programs are affordable).

As technology changes, so does the communicative potential (for good or evil) of the medium. In the early days of television, before videotape, programs were either aired live or captured on film, which was more expensive and laborious than tape. Satellites have changed international news reporting and international diplomacy as well. Similarly, lightweight, hand-held video cameras have expanded the limits of what images can be captured, and they have had an important effect on television drama, news, sports and rock videos. They have also had a remarkable impact on home photography, as evident in the phenomenal success of some of the home-video programs on the commercial networks.

But the actual, day-to-day limits of television are determined by the program producers, distributors, retailers, regulators, viewers and critics, who often battle for financial power, personal glory and altruistic service—to mention just a few motives.

Neither the advocates nor the critics of television are clear about the real or artificial limits of the medium, because they fail to examine the various characteristics of both the technology and the institution. As a result, many critics suggest that television is inherently evil, when in fact they should say that only the social institution is evil. More precisely, the critics should say that television's use is often governed by evil values, beliefs and practices.

For example, at one point Owens says that television cannot communicate the authentic gospel because it is "in the pile of rejected possibilities that the cameraman leaves behind when he packs up his gear and leaves town." Yet she also says that the camera "cannot hope to catch" that with which "the gospel is concerned."[41] In the first example, she blames television's inability to communicate the gospel on the institution of television—the camera operator who decides not to make spiritually authentic programming. In the second quote, she criticizes the technology—the camera itself—for the medium's lack of spiritual content.

Similarly, Postman's popular critique confuses technology and social institution. He argues throughout the book that television inevitably turns programs into mere amusement, thereby trivializing such important subjects as politics, education and religion. The problem, he says, is the technology's emphasis of image over printed words; only words can present propositions that can be discussed and debated thoughtfully.

Yet in the last few pages of the book Postman does not take his own advice to heart. He says the "solution" to the TV problem must be found in teaching people "how to view"—presumably a project for the schools. In other words, Postman locates the problem in the technology and the solution in the social institution. He can't have it both ways.[42]

As a communications technology, television is *not* culturally neutral. On this point Postman and others are correct. The tube will always, by its very nature, communicate differently from the way other technologies do. Like all communications technologies, it becomes part of the message and therefore part of the culture. Each communications technology influences society and individuals in particular ways—no matter how it is used.

Compared with printed communication, for example, television may indeed be an inferior medium for philosophical argument, as Postman suggests. But that does not mean the technology of television is inherently evil. Instead, it suggests that producers and viewers should seek the best uses of the technology for the good of humankind and the glory of God.

Certainly every new medium can be a threat to existing ones. If people watch more television, they might read less. However, as Tony Lentz has argued, when there is a "dynamic tension" between old and new media, there can also be a "flowering of culture."[43] Every communications medium offers some potential contribution to fulfilling the Cultural Mandate and the Great Commission. This means teaching producers how to create programs responsibly and instructing viewers on how to use them wisely. This book seeks to help Christians do both.

The celebrators of television fail to distinguish between television as a technology and television as an institution. When the televangelists speak of the redeeming power of the medium, they are often oblivious to how the technology influences culture and how the social institution shapes the message. For example, Armstrong's rhapsodies about the impending global church fail to take into account all the conflicting messages broadcast by hundreds and perhaps thousands of televangelists around the world. Institutionally speaking, contemporary televangelism is a cacophony of disparate voices and images produced with widely different standards and values—not to mention ethics. If people are saved through television, it will not be due to the power of the technology as much as to the authenticity of the message, and that is controlled largely by the social institutions that make the programs.

The critics of television often lay all the blame for the medium's evils on the institution, while the advocates of television praise the technology. Both groups say they are talking about "television." As I argue throughout this book, the process of "redeeming" television requires Christians to make some difficult judgments about the interaction of technology and social institution.

Conclusion

As Edward Carnell once wrote, it is "unrealistic to believe that TV will be either all good or all bad."[44] It's too easy for each of us to veer toward either hopeful veneration or cynical condemnation of television, thereby avoiding careful and spiritually discerning assessment.

Like the surrounding culture, the church badly needs a map to guide it through the maze of conflicting ideas and perspectives about the medium of television. The map must take into account the tension we all feel between the wondrous grace of God and the pernicious evil of the Fall—especially our own sinfulness.

In the remaining chapters I will offer a map of television, looking for ways of redeeming it. In the process, I hope we can get beyond and above the confusion of the contemporary debate.

As you read my "map," keep in mind three important ideas.

First, God gave communication not only for directly preaching the gospel but also for fulfilling the Cultural Mandate. The complete redemption of television requires us to consider the medium's potential for creating, maintaining and transforming culture for the glory of God. Evangelistic programs are not enough. News, drama, entertainment, sports—all fall within the scope of the Cultural Mandate. And all of them have an evangelistic dimension in the sense that they establish the cultural context for the gospel.

Communication and culture point simultaneously to the Great Commission and the Cultural Mandate. We need not feel guilty about watching programs that don't preach the gospel. But we cannot deny the evangelistic aspect of all communication, including the most apparently secular programming. Culture is never a neutral conduit for the gospel.

Second, redemptive television is part of our vocation as image-bearers of God. All humans, whether they work in television or not, are called by God to use the medium responsibly. The freedom to view television carries with it a vocational accountability before God. We are meant to be responsible caretakers of the creation, even as we watch the tube.

The commercial networks, some televangelists and millions of viewers seem to think they are merely in business for themselves. Television belongs to God. We are to use it wisely and discerningly.

Third, the medium of television is a combination of technology and social institution. In order to redeem it, we must consider both the inherent nature of the technology and the human nature of the institutions that use it. Rather than blasting away at what we do not like about television, or using it thoughtlessly, we need to consider what beneficial role it can play in society and in our personal lives.

Let's be realistic: Television cannot be all things to all people. It has its own limits as well as its own potential for good.

Ironically, I learned these lessons from a television interview with one of the medium's most articulate Christian critics, Malcolm Muggeridge. More than most broadcasters, Muggeridge saw the cultural and spiritual dangers in the use of television. His criticisms are a good antidote to the unreflective triumphalism of much Christian TV. But we still need a map that will help us find our way past Muggeridge's despair as well as past some televangelists' inflated optimism.

2
TALES ON
THE TUBE /

How TV Shapes Culture by Telling Stories

In their greed these teachers will exploit you with stories they have made up.
(2 PET 2:3)

Y EARS AGO I STOOD NERVOUSLY BEFORE A GROUP OF SEVERAL THOUSAND teenagers. As the main speaker at a youth event, I was responsible for delivering a multimedia presentation about the effects of popular entertainment on adolescents. When the spotlight came on me, and the auditorium lights went black, I couldn't see a thing, except for my script on the podium in front of me. So I began reading.

The next thing I knew, a few pennies sailed by my head, landed on the platform behind me, and rolled around there until they came to rest. Then I heard a few soda cans (or was it beer?) popping open in the darkness. Paper airplanes floated down from the bleachers.

I had nearly lost the group, so I gave the signal for the first video clip. Some young people cheered. Others jeered. Hundreds of them started singing along with the clip; the bedlam escalated. The presentation had turned into a struggle between an irrelevant professor and a playfully hostile audience.

After that experience, I knew I had to change the presentation. I began adding personal stories as illustrations. The more I personalized the presentation and the more I used story illustrations, the greater the audience's attentiveness. Whenever I tried to teach an important point without using a story, I would begin losing the audience.

I found that stories, especially personal narratives, have an incredible power to capture listeners and to gain their respect. My listeners realized that I had lived the lessons—and they enjoyed the tales.

My experience with youth held true with adults and children too. This changed my teaching and lecturing style. Instead of simply trying to teach ideas, more and more I found myself teaching through experience—my experience and the experiences of other people, from textbook authors to filmmakers. In other words, I linked communication to narrative experience, not just to abstract ideas or important lessons and themes.

Most effective communicators are savvy storytellers—people such as Tony Campolo and Bill Cosby immediately come to mind. Storytelling is a gift, but it is also an art and a business. The television industry learned about storytelling from stage, film and radio. It had to, because drama is the performance of stories.

Our attitudes about storytelling are often shaped by our views about entertainment, which is the major purpose of public storytelling in North America. Christians are sometimes torn between two opposing views of popular drama. One view takes its cue from the broader culture, assuming that television is "just entertainment" as long as it doesn't overtly challenge the faith. I hear this argument all the time from Christian college students reared in permissive families that consumed liberal doses of media.

The other view takes its cue from the more fundamentalistic religious subculture and looks upon all the "worldly arts" with suspicion. Entertainment is frowned upon except when it clearly teaches a Christian message. Christian music is acceptable, secular music is not. Christian TV is good, nonreligious TV is evil. In some cases, the media themselves are rejected as evil.

It seems to me that both of these views of entertainment are wrong. And they lead to Christian misuse of television.

Even though entertainment is not "just entertainment," neither is it inherently evil. Entertainment is the process of making and consuming leisure-time products, especially media products. The moral nature of that process depends both on *what* we consume and *how* we consume it. The full goodness or evil of a program is not inherently in the show. This is especially true of dramatic stories.

This chapter is a primer on stories for everyone who wants to communicate more effectively—especially through television. Both producers and viewers will find the examples and illustrations helpful.

Television, as North America's major public storyteller, is an entertainment giant. Although the tube is especially popular among children, older and elderly adults, and the poor and less affluent, including minorities, it is truly a mass medium used by more than two-thirds of the homes in the United States every evening.[1] In Canada, too, television is an enormously popular storyteller, as it is increasingly around the world.[2] Its tales often dwarf those of our local communities, churches and families. We cannot fully evaluate the tube without assessing how we use the medium's stories, not just how the media professionals use television.

From the creation perspective, all stories that help us to serve God and humankind, to take care of and celebrate creation, are worthy of our time and energy. We have much work to do and much in life to enjoy. If stories help us to accomplish our God-given tasks and to delight in his creation, we should embrace them enthusiastically and joyfully. If they lead us from these activities, we should be wary of them. We are not called to worship stories or to give them up, but to use them for the furtherance of the kingdom and to the glory of God.

Among human gifts and talents, storytelling is clearly an important one. Television is a significant way to exercise that gift.

There are four major ways we can use stories for the glory of God: amusement, instruction, confirmation and illumination. Each one is important, but each has its dangers. Television can contribute wonderfully to each use of story. It can thereby make our lives more meaningful and enhance the culture that surrounds us. Moreover, the tube can use these approaches to stories to help communicate Christian culture and witness to the world the lordship of Jesus Christ.

Amusement: Relieving Boredom
Amusement is probably the most basic and universal function of narrative. No one knows this better than television writers and producers.

As human beings create culture, they tell stories partly to relieve boredom.

Amusing narratives occupy the mind, but not with deep or reflective thought. Most amusement is comedic rather than tragic; it is far easier to amuse through humorous tales than through serious, foreboding stories. Some types of humor, such as slapstick comedies and silly cartoons, serve this function easily and almost effortlessly. From its earliest days, the tube was loaded with vaudevillian variety programs, but soon the situation comedy became the most characteristic form of television drama. David Marc has even suggested that television is America's "jester," providing lighthearted humor not for a king's court but for a whole nation.[3]

This function of narrative has been attacked repeatedly by intellectuals and scholars. Neil Postman, for example, argues that the tube's inherent bias toward amusement is destroying the fabric of public discourse in America. In his view, practically all aspects of life, including politics, are trivialized by television.[4]

Postman believes that television naturally promotes poor thinking, what "Tonight Show" founder Steve Allen calls "dumbth."[5] Books promote learning, thought and rationality, while television produces passive, thoughtless human beings mired in trivial amusement. "The problem," writes Postman, "does not reside in *what* people watch. The problem is in *that* we watch."[6]

In one sense critics such as Postman are right. Little on commercial television encourages deep thought, or even moderate reflection. Viewing television is not normally an intellectually stimulating activity. Most viewers are quite passive—both mentally and physically. Programs such as "Gilligan's Island" and "The Beverly Hillbillies," one of the most popular shows of all time, rarely stimulate viewers' intellectual curiosity or foster serious public discourse about anything. Neither did David Lynch's "Twin Peaks" or the nutty "Family Matters."

Nevertheless, there is nothing wrong with amusement per se. In fact, all of us find amusement in some activity. As anyone who travels on airplanes knows, superficial novels, gossip magazines and television-style newspapers are the mainstay of the bored traveler. Airport gift shops are loaded with shallow reading material.

The fact is that amusing tales long predated the electronic media. From religious ritual to the Shakespearean theater to modern magazines, amusement has always been one of the functions of dramatic and literary storytelling. Even when stories were meant to engage the mind, they had first to be entertaining in order to attract audiences.

Humans have a basic need for relief from the day-to-day grind of life; consequently, we seek out amusing diversions. Amusement is a legitimate function of narrative, and to expect all stories to meet rigorous intellectual standards is ludicrous and arrogant.[7]

Humankind does not live by rational discourse or literary prowess alone. Nor does it live by work alone. Many television critics wrongly place mental work above any other human activity. A life filled with nothing but intellectual activity would be just as unbalanced as one filled with nothing but comedy viewing. That's why even television's critics often watch amusing films or read amusing books.

The Dangers of Amusement

But not all amusement is equal—morally, artistically or spiritually. Amusement is not always good for us.

Television is never merely an escape from the real world. It's always an entry into another world created by someone, usually for financial reasons. Situation comedies, for example, amuse in very different ways. "The Cosby Show," "M*A*S*H," "Cheers" and "All in the Family" were very different forms of amusement; they varied in seriousness, satirical content, predictability and simple diversionary power. While humor often amuses, it can also celebrate our humanness or mock people. Humor can be at other people's expense or at our own. It can be crude or clean, prideful or self-deprecating.

Perhaps the most chilling condemnation of contemporary television is the seemingly insatiable appetite it creates for amusement of all kinds without regard for social or moral benefits. As we scan the channels with remote-control switches, we may not be looking for the best amusement, only the show that diverts us most effectively from the problems of the day or the boredom of the moment.

Stories can amuse for many reasons—witty dialog, interesting characters and the like. However, the most compelling aspect of narratives as amusement is probably the way they catch viewers in anticipation of what will happen next. Soap operas successfully use anticipation to hook audiences for years. As viewers get vicariously involved in the televised tale, they are diverted from the stresses in their own lives and caught up in the experiences of fictional characters. The desire to know how a story ends frequently gets in the way of making discerning judgments about whether a story is worthwhile.

Because of the sheer quantity and availability of televised amusement, we are not often inclined to be discerning about our choices of programming. Relief from boredom becomes the sole criterion for some viewers, as well as for the television industry. As an institution, commercial television has fostered the mistaken belief that amusement is only a matter of personal taste and individual freedom. While some intellectuals would encourage us to abandon amusement, the commercial television industry would have us define amusement as the most worthwhile human activity.

The entire television ratings business encourages this myopic view of amusement; audience ratings are generally little more than measures of which shows were most amusing. Ratings indicate nothing about the source or quality of that amusement. "Night Court" may be rated higher than "Full House," but the latter offers a much more wholesome amusement.

When amusement becomes the guiding principle behind all of our leisure, we might indeed turn into the kinds of mindless citizens depicted by Postman. In fact, too much of anything, including reading and study, could keep us from dealing with important duties and responsibilities, as well as from seeking deeper sources of delight and joy. Couch potatoes who spend all their free time in front of the tube are missing out on the joys of warm relationships with friends and relatives. They also lose the benefits of other art forms. Like any single pastime, unbridled television viewing can keep us from important activities such as prayer, conversation and reading. In the television age, we can easily elevate amusement so much that it takes on a godly status as the most important cultural activity.

Perhaps there is something about modern industrial society that has made amusement more important in people's lives. In late twentieth-century America, television seems to be the drug for most people addicted to amusement.[8] Television is our escape from the routine of life—our diversion from alienation, the stress of work, the frustrations of bureaucracy and the lack of meaningful relationships. Perhaps the stress of sour personal relationships leads people to the tube as a means of temporarily alleviating emotional suffering. And that's not all bad if we do not take it to an extreme.

It's also possible that the electronic media are replacing other forms of amusement—especially oral storytelling by families, friends, neighborhood groups and even churches. Maybe the professional storytelling of the electronic media is more attractive and exciting than people's own narratives; we watch television instead of amusing ourselves with personal stories. Certainly folktales are not widely told in America. Even nursery rhymes and other children's stories have been mostly replaced with televised cartoons and action programs.

Except for gossip, we don't tell each other many stories, and young people today only infrequently hear tales about previous generations of their own families. Television might not be the major culprit, but television's stories are so slickly produced and so readily available that we apparently find them more appealing.

Unfortunately, mass-produced, electronic narratives do not arise out of our own community and family life. By focusing on the tube instead of our own stories, we lose continuity with our past. Moreover, we endanger all social institutions that depend upon generational continuity, including the family and the church.

Instruction: Teaching Lessons, Rules or Precepts

The fact that stories can teach lessons is obvious to the Christian community. Christians use biblical parables, personal testimonies, moralistic fiction and evangelical-conversion films to teach each other about the faith. Some of the most inspiring Protestant preachers are fine storytellers, in their ability both to use engaging illustrations and to retell biblical events. It's probably fair to say that many of the most popular preachers are usually the best storytellers.

Biblical history is a rich reservoir of lessons about God and humankind. If the Bible were merely a set of laws or a catechism book, people would find it far less interesting and convicting.

Television is one of the major educators of modern society. As William Fore once put it, the mass media are "classrooms without walls."[9] Television is not usually a formal educator, such as with telecourses offered by colleges, but an informal, implicit, often unconscious educator. While the commercial television industry claims to be merely in the entertainment business, it is inherently in the instructional business as well. Viewers *learn* from the tube, even if they aren't interested in an education. Televised stories teach many things.

But instructional storytelling is complex. Some of the instruction is the result of simple role modeling or imitation; we learn styles of dress, linguistic patterns, vocal inflections and even specific words. The Fonz and Bart, in

"Happy Days" and "The Simpsons," were influential teachers of this sort.

Viewers even learn how to think and feel from the attitudes and values expressed on the tube, as well as its stereotypes of social classes. One of Bill Cosby's goals was to teach whites that blacks were not lazy ghetto folk, while Roseanne Barr attempted to show that it's acceptable to be an imperfect family.

The success of "The Cosby Show" in the 1980s probably did more than anything else to convince the white-dominated industry to make programs about blacks with implicitly instructional messages. African-American characters on programs like "A Different World," "Family Matters," "Gabriel's Fire" and "Fresh Prince of Bel Air" tended to be multidimensional people instead of shallow stereotypes. As a result, white viewers learned about blacks, and blacks were taught that they need not live out whites' stereotypes of them.

"A Different World," for example, portrayed black college students instead of criminals or uneducated, unemployed people. It even appears that "A Different World" significantly increased interest in black colleges.[10]

The point is that television teaches on many levels, not just through a show's overt message, and there is no way to determine precisely what one learns from the images and sounds emanating from the tube. Different audiences can interpret programs quite differently.

For example, working-class women are more critical of soap operas than middle-class female viewers are. Both groups learn a conception of middle-class reality from the soaps, but working-class women are more likely to reject the portrayals because they do not fit their expectations of middle-class life.[11]

Television audiences are not totally passive students. They bring their hopes and fears—all of their past experience—to the set. Moreover, both our conscious and our subconscious are involved in watching television. Just as a preacher communicates through mannerisms and style, as well as the words of the sermon, television's instruction goes beyond the dialog of characters and the themes of the stories. Some programs are explicitly moralistic, guiding the viewer toward a particular message or lesson. But most of them are not nearly so simple to interpret and evaluate; they communicate *implicitly* in the guise of mere entertainment.

Television instructs most when viewers know little about the subject matter and do not hold strong opinions about it. This is why American-made shows can be so influential in shaping other cultures' views of North Americans. The tube may be the only "direct" contact some viewers have with American citizens. Millions of people around the world have "learned" about Americans from "Dallas." Although the series is interpreted differently in various cultures, it still establishes an image of Americans as a people and as individual types of personalities—especially for viewers who do not meet Americans personally.[12] The sensibility of a program, its overall attitude toward life, is one of the most influential modes of instruction. The humorous put-downs of TV sitcoms shape peoples' real-life dialog. Teenagers have been particularly vulnerable to this kind of implicit instruction, transferring put-down humor from the tube and movies to their daily interaction with each other. After all, they have few examples of kindness and compassion on television comedy, which is dominated by satirical humor.

Today's adults grew up in an era when the tube's instructional sensibility was far more obvious. Early situation comedies were clearly moralistic programs; in the 1950s, they generally told stories with clear, unambiguous endings. They were secular parables with simple, widely held moral precepts: "Tell the truth," "be patient," "don't be a hypocrite," and so on.

Such shows did not necessarily call for moral change as much as display the fruits of personal morality. Indeed, they were like popular morality plays of the nation's civil religion. "Leave It to Beaver," "Bachelor Father" and "My Three Sons" will always be remembered for their moral innocence and instructional simplicity; they were a far cry from the more cynical and ambiguous "All in the Family" or even "Roseanne."

Hour-long dramas such as "Bonanza," "Gunsmoke" and "Wagon Train" were also moralistic tales that instructed Americans about right and wrong, good and bad. Compared with today's more ambiguous morality plays like "L.A. Law" or "Cheers," the older programs offered clear moral guidance.

Nevertheless, even morally chaotic series like daytime soap operas have their educational impact. This genre, more than the others, daily plunges viewers into private lives and sensitive topics. Soaps connect "with basic human concerns, explaining complex social phenomena, [and] providing categories for thought and moral precepts to live by." As one viewer put it, "We can relate to the situations and sometimes sort . . . our own problems out through listening and doing what the character portrayed has done."[13]

Among the things we have learned from the tube is an idealized, romanticized image of physicians. In his book on how prime-time TV dramas have portrayed doctors and medical institutions, Joseph Turow concluded that "TV fiction can have a major effect on the perception of, and clout of, an institution in society." As the networks and production companies portrayed doctors, they shaped public views of the profession. "Most of the time," says Turow, "the relationship between TV producers and the doctors has been symbiotic, a 'you scratch my back, I'll scratch yours' approach that has served the interests of both parties. Physicians have, by and large, received favorable treatment from producers. For their part, program producers have had convenient places to turn for inexpensive advice as well as a patina of credibility regarding accuracy."[14]

More recent series such as "St. Elsewhere" were not as sanitized as the old "Ben Casey," "Dr. Kildare" and "Marcus Welby, M.D.," but there is little doubt that most Americans have learned more about what doctors do from the tube than they did from any other medium—including their own infrequent conversations with their physician.

Instructing Kids and Teens

Young children spend a phenomenal amount of time in front of the set—roughly twenty-eight hours weekly. The average child will have spent more than six thousand hours watching television before starting formal schooling, and about twenty-three thousand hours by the end of high school. This represents more time than children spend in school during their social-learning years.

Moreover, the preschool child spends about five hours weekly watching television commercials—over a thousand advertisements each week. Before a child enters school, he will have been "instructed" by the advertising industry in twenty-four thousand commercials. By high school the figure is over a million commercials.[15] In other words, children are heavily propagandized by trained "educators" long before they begin receiving formal schooling. It is probably fair to say that television is the major teacher of values beyond the home. Parents face an enormous task of educating their children about this implicit mass-media instruction.

During the 1980s the commercial television industry learned that one of the most powerful ways of attracting preschool and young school-age children was to provide stories with attractive role models. Program producers discovered an old truth: Young children like to learn through imitation. The industry realized that role models both boosted audience ratings and provided new markets for toys and other products associated with the programs. Shows such as "He-Man and the Masters of the Universe" and "G.I. Joe" made millions of dollars for television stations, program distributors and toymakers. So did the popular Teenage Mutant Ninja Turtles. Tens of millions of American children learned about these televisual role models and then integrated the toys into their own imaginative play, acting out the lessons and behaviors aired on the tube.

Soon virtually all commercial children's programs, as well as many public television shows, were marketing a vast array of television-related lunchboxes, underwear, toothbrushes, flashlights, backpacks and almost every imaginable child product. The goal was clear: to teach kids to be like the role models who acted out various precepts and lessons on the tube. In the simple moral universe of the young child, this was powerful programming that heightened the tube's nurturing role in his life.[16]

Teenagers also learn from the tube, but not just through role models. According to a study conducted by the Department of Education, eighth-grade students spend nearly twenty-two hours weekly watching television, compared with about five and a half hours doing homework and two hours of outside reading.[17] From "American Bandstand" to MTV, television taught adolescents how to solve their teenage "problems." Caught in a rapidly changing society, adolescents looked to media such as television for ways to dress, talk, comb their hair, and so forth.

One might define adolescence in North America as the time between childhood and adulthood when young people are unsure about who they are (identity) and confused about how to establish close emotional relationships (intimacy). Television, through the VCR and teen flicks especially, offers adolescents attractive ways to meet these needs. The media become quasi-educational institutions, giving guidance for profit.[18]

Rock stars like Bruce Springsteen and Madonna were teachers in this new mass-mediated classroom. If a female adolescent could look like Madonna, she believed, it might be possible to achieve popularity with other girls while enhancing her attractiveness with boys. Such visual lessons were almost always tied to adolescent sexuality; natural hormonal changes made teenagers eager students of gender images on television and in films. MTV, more than any other program or network, created an entire visual and aural atmosphere for adolescents to learn rules of identity and intimacy—rules that would also lead American youth to buy records and videos as well as attend concerts and purchase advertised products.[19]

This quasi-educational use of TV by teens helps explain why there are so many references to sexuality in adolescent-oriented media. Researcher Bradley Greenberg estimates that teens are exposed to up to four thousand sexual references yearly on television and film. These include at least fourteen hundred references to sex on prime-time TV and another thousand on daytime soaps. One hour daily of MTV and one R-rated film per month contribute sixteen hundred sexual references.[20]

A survey of adolescents in conservative denominations found that friends and movies were their main sources of information about sex.[21] Neil Postman and his colleagues found in a study of beer commercials in 1987 that advertisements developed the myth of masculinity associated with a boy's passage into manhood. Advertisements don't just sell beer; they peddle masculinity and adulthood.[22]

More and more, television (including televised movies) competes as a teacher with traditional social institutions whose job it is to nurture children and teens—schoolteachers, parents and pastors. Cable TV and the VCR have given the tube even greater power to teach youth. Clearly, the Christian community cannot ignore these fundamental social changes which put the school, church and family on the defensive.

More Instruction Is Less

Ironically, the most explicitly instructional narratives may actually teach the least. Most viewers don't want to be preached at by the tube. Moralistic tales usually turn audiences off—except for those viewers who already agree with their message. This is why Christian television, which tends to emphasize teaching and preaching, is generally unable to attract many non-Christian viewers. It is also why MTV, which purports to be merely entertainment, is actually a powerful teacher. All televisual stories teach, but the most implicit teachers are often the most powerful—unless the viewer wants explicit instruction.

The power of implicitly instructional stories usually lies in the way they simplify and stereotype. Such stories don't ask questions or portray ambiguity, but offer simplicity and certainty. Commercials are the most obvious examples. They teach us how to achieve unquestioned ends—a "perfect" body, material happiness and the admiration of friends.

Similarly, TV shows that clearly distinguish between good and bad characters have usually been the most popular ones on the air. Westerns succeeded in the 1950s and 1960s by packaging and repackaging moralistic tales about the good guys winning the battle with the bad guys.

More instruction is usually less. Heavy-handed teaching tends to drive away all viewers except those who are looking for a lesson. On TV this can be financially deadly, since the vast majority of viewers want entertainment. Teaching on the tube becomes a powerfully subversive activity when it is anchored in narratives that purport to entertain.

Confirmation: Validating Existing Beliefs and Values

Throughout history, tales have been used to express the fundamental beliefs of various cultures.[23] Even in the modern postindustrial world this mythopoetic (mythmaking) function of narrative is vibrant and influential. Stories can powerfully confirm what people believe or want to believe about life. The answer to the "chicken-or-egg" argument is "both": television both changes the culture through instruction and maintains the culture by confirmation. It teaches new values and beliefs while affirming old ones. In other words, the tube instigates change *and* embraces stability.

Deciding when the medium is teaching something new or merely reaffirming existing values and beliefs is often not easy. For example, Ella Taylor found in a study of how television portrays families that the tube "feeds off itself and other media, and in this way its images both echo and participate in the shaping of cultural trends."[24]

Television is the most significant medium for mythologic storytelling in contemporary American society. Like the bard or the epic poet, television transmits narratives from generation to generation.[25] Partly because it has to attract large audiences, commercial television tends to use stories that confirm what Americans already value and believe—or at least what they *want* to believe—about themselves. This makes popular television an effective propaganda vehicle in North America. As Jacques Ellul has argued, the most potent propaganda tells people what they want to hear and believe.[26]

The confirmatory power of television results from its narrative formulas. "The Cosby Show," for example, was a variation on the moralistic comedies of the 1950s and 1960s. It confirmed the same kinds of traditional family values, but frequently with a serious twist that made its stories a bit more realistic. The mythic genius of popular television is its remarkable capacity to reformulate the same *kinds* of stories in myriad ways—amusing with novel material while confirming old values and beliefs via established genres.[27] Each of the tube's major genres, from the sitcom to the western, the detective show and the newer ensemble series such as "L.A. Law," is essentially confirmatory.

Because they have to attract audiences for advertisers, commercial series generally communicate themes that millions of viewers feel affirm their own beliefs. Shows can offend with particular scenes that are distasteful to audiences, but usually they cannot challenge viewers' *basic* beliefs. Soap operas, for instance, might offend some viewers' moral sensibilities; at the same time, though, they "largely preserve and reconstitute the gendered status quo" and encourage viewers "to yearn for perfect love in a mythic community . . . where family and community still count, and where a moral and social order provide stability." Regardless of how much things seem to change on the soaps, the genre's melodramatic world still affirms the things that viewers want to believe in.[28]

Commercial television, then, tends to validate particular American beliefs by repeating established formulas.[29] This is why almost all programs are so predictable. Most viewers do not want their beliefs challenged, so the tube combines novelty and predictability to attract viewers to new versions of old stories. As David Marc has suggested, "While individual episodes—their plots and climaxes—are rarely memorable (though often remembered), cosmologies cannot fail to be rich for those viewers who have shared so many hours in their construction."[30] In other words, the tube has generally confirmed popular world views, sentiments and attitudes.

It is the mythic, confirmatory nature of much television programming that makes it a "living tradition" of its time.[31] Every decade has its popular shows that somehow manage to capture something of the spirit, sensibility and values of the period. In the 1970s what could have been more emblematic than "All in the Family," which reflected (and reinforced) the generational tensions and conflicts among Americans at that time? And what would confirm the optimism of the Reagan years more unabashedly than the Huxtables of "The Cosby Show," where people really did love each other, they could actually get along, and, despite the conflicts of the day, everything looked rosy?[32]

Mark Crispin Miller goes so far as to suggest that "The Cosby Show" was popular with whites "in part because whites are just as worried about blacks as they have always been." He suggests that the program reassuringly told whites that blacks were a "threat contained," thus negating the "possibility of black violence with lunatic fantasies of containment."[33] This power to confirm stereotypes through images may be one of the most significant impacts of the tube on society, especially in a period of globalization, when so many ethnic and racial groups are exposed to images of other groups on television.[34]

The tube's confirmatory power should not surprise Christians, who have long used art and story to nurture faith communities. Since Christ's birth, much of the West's art has been produced either by the church or by individual artists influenced by the Christian faith.

As Nicholas Wolterstorff suggests, confirmation has probably been the most pervasive benefit of art. "Over and over when surveying representational art," he writes, "we are confronted with the obvious fact that the artist is not merely projecting a world which has caught his private fancy, but a world true in significant respects to what his community believes to be real and important. Since in most communities it is the religion of the people which above all is important in their lives, this implies that much of the world's representational art is explicitly *religious* art."[35]

Television as Religion
In the contemporary world, this mythopoetic function of art, especially storytelling, shifted to secular institutions, including commercial television. The new storytellers were no longer interested in religion per se, but in making money by telling stories that sometimes resonated with quasi-religious beliefs. Commercial bards were driven by the marketplace to create confirmatory narratives that are increasingly exported around the world.

In Mali, West Africa, traditional oral storytellers, called *griots,* are now challenged by television. As one of the most famous *griots* puts it, "Children are more interested in watching television than listening to a traditional storyteller. . . . It is very difficult to prevent children from watching television. Now is their time. Now is television's time." As a result, young Malians will know less about their own cultural past and will not as likely learn how to serve their country by imitating their forebears.[36]

In other words, commercial television sometimes functions like the sacred stories of a religion, confirming the myths of secular society. On one level, television reflects an "American middle-class morality" that was crystallized in the nondenominational evangelicalism of the early nineteenth century, says Victor Lielz. Even a show such as "M*A*S*H" displays such historic middle-class virtues as willingness to work hard, creatively and selflessly; devotion to duty, comrades and nation; technical skill; personal integrity; and compassion.[37]

On a deeper, nearly religious level, the tube's basic values include the following: (1) good triumphs over evil, (2) evil people cause evil, (3) evil can be eliminated by eliminating these evil people; and (4) society can be redeemed by the good works of moral individuals.[38] Not all American television drama neatly fits into this "theology," but the great bulk of it does. The same basic themes appear repeatedly in Westerns, comedies, detective programs, children's cartoons and made-for-TV films.

The tube's confirmatory power rests in the creative ways it reformulates old myths in new packages. As J. Fred MacDonald has argued, the western died on television because it did not adapt to the changing social values of the 1960s and 1970s. The genre became "incompatible with a civilization where the flow of events . . . forced a reevaluation of the innocence and satisfaction with which most Americans had accepted the functioning of their society."

Eventually the western was replaced with the more morally ambiguous detective show, which used the new urban gunslinger to bring law and order. Detectives often broke the law in order to uphold it, thus confirming society's faith in the triumph of good over evil, but in a more cynical context. MacDonald concludes that "life for the majority of Americans remains an urban ambiguity—a world of compromises with city life and modern technologies in which workable answers are not readily forthcoming."[39]

Imagine what would happen to audience ratings if all the commercial networks suddenly transformed all prime-time programs into tragedies: Cliff Huxtable of "The Cosby Show" accidentally shoots his wife fatally in the head; Starship *Enterprise* and its entire crew are blown to pieces on "Star Trek"; the neighborhood gang on "Cheers" is murdered by poison put in a beer keg by a maniacal terrorist. American viewers do not want such unpleasant stories, which would challenge the secular theology of prevailing social myths.[40]

Commercial television can survive by offering conflicting myths as long as different social groups can interpret the stories in ways that support their own perspectives. Julie D'Acci believes this happened with "Cagney and Lacey," which provided different—even conflicting—portrayals of the role of women in work and family.[41] Her findings make a lot of sense, particularly for shows with deeper characters and more adult themes. For example, "thirtysomething" often walked the line between respecting tradition and embracing unbridled personal freedom. It could be enjoyed and interpreted either way.

Television, then, is in the business of quasi-religious confirmation. It competes with parents, teachers and 'pastors in this regard; these traditional authority figures have their own tales to tell, but television's stories seem far more entertaining to millions of viewers. As a result, the tube becomes the bard of choice, overshadowing other media of confirmation. Television is given (and takes) the power to validate the cultural status quo.

Of course, the technology of television could be used to point viewers specifically to religion, or even to a particular faith or denomination. Yet the tube, which is designed to reach as broad an audience as possible, is decidedly nonsectarian. Hardly ever do TV characters "read books or invoke religious beliefs in order to settle arguments. When a problem arises on a sitcom it is solved pragmatically. The standards invoked are general and non-sectarian."[42] Writers of television series appeal to most basic American myths, not to particularly Judeo-Christian ones.

Television programming and viewers are not a religion per se, but function in a quasi-religious manner. It would be wrong to conclude that television is "America's cultural religion," as John Wiley Nelson does.[43]

First, American culture is remarkably diverse. On the most basic cultural level there are common beliefs, but even many of these may not even be distinctly American. Nor are they necessarily in total conflict with sectarian beliefs promulgated and celebrated by distinctly religious groups, including evangelicals. Christians, too, believe that things will work out for good in the end—but for those who love the Lord and are called by him. It might be that the "secular" ideas of television, such as the belief in progress, are cut from the vines of Christianity in the Western world.

Second, referring to popular art, including most television programming, as religious artifacts is far too broad a notion of what constitutes genuine religion. In one sense all of life is indeed religion. Everything we do is in response to God's creation. But not all humans even attempt to live according to a system of belief that recognizes God. Religion requires some belief in a transcendent Creator before the creation itself can be part of religion.

Third, popular art does far more than confirm belief. Certainly this is a significant function of television, but there are others, including amusement and instruction. Consider the case of "Marcus Welby, M.D." Nelson writes, "When Dr. Welby accomplishes another 'cure,' we come away believers again."[44] Believers in what? Medicine? Science? Common sense? Fatherly doctors?

Only at the most general level of belief—that human beings can solve their own problems—is there a common myth. This basic form of secular humanism is faith in humankind without the necessity of God. To the secular mind televisual mythology may be "religion," but not to the Christian. Certainly television can *function* like religion, but it is not a religion.

The Problem of Cultural Pluralism
Although we might not agree with the popular myths corroborated by commercial broadcasting, the confirmatory role of story is certainly legitimate. Old verities need to be passed down from generation to generation if society is to survive. We must also accept the fact that citizens in a pluralistic society such as the United States or Great Britain must share some beliefs if they are to work together as a nation for the good of everyone.

Even when we are most outraged by the naive and misguided "faith in ourselves" confirmed by television, we must recognize the need for a national public life based at least partly on common beliefs. Television can legitimately confirm democratic values such as the importance of individual life, the superiority of democratic political systems and the individual's responsibility in society. Without public stories to affirm these values, democracy could not likely survive, and the nation would probably fall into self-serving individualism, social chaos and cultural disintegration.

The Cultural Mandate requires us to use confirmatory stories for the maintenance of society. The Great Commission requires us to infuse those stories with the grace of God.

However, in a pluralistic society a story that confirms what some people believe can just as easily challenge what others believe. When public advocacy or pressure groups take on the networks, they are reflecting such conflicts over confirmation.

When the made-for-TV film "The Day After" appeared on network television during the early Reagan years, many political conservatives interpreted it as liberal propaganda.[45] Set in the Midwest during the hours after a nuclear attack, the film offered viewers gruesome scenes of people fighting to survive. Many liberals hailed the film's realistic portrayal of a postbomb scenario, even if they believed that it was unrealistic in suggesting that many people might survive a nuclear attack. They argued that the program encouraged more intelligent public debate about the effects of a nuclear attack on the United States.

Disputes about television programming are often battles over whose values and beliefs should be confirmed and whose should be challenged. Christians should participate knowledgeably in these kinds of disputes, affirming democratic values on the one hand and asserting their own right to tell their religious story on the other.

In addition to amusing and instructing, then, television can publicly ratify already-held beliefs while challenging minority beliefs, including theistic beliefs. As Edward Carnell once put it, television threatens the "secularization of our culture" because of the medium's "efficiency . . . to mirror the world in the home."[46] Action shows do it by reaffirming the long-held belief that good triumphs over evil. Situation comedies, especially in the early days, did it by confirming the goodness of human beings and the seemingly eternal nature of the patriarchal family, as well as the myth that all things will work together for good. Even rock videos confirm young people's sense of the joys and pains of adolescence.

Driven by the economic necessity to attract large, heterogeneous audiences, commercial television often presents narratives that confirm the widely held beliefs of the culture. In this sense, popular drama serves the mythopoetic function long held by religious art and liturgy. Hollywood producers are the new acolytes who serve up the latest icons and stories for the congregation of viewers. Viewers are the new congregants.

Illumination: Revealing and Clarifying Aspects of Life
Finally, televisual narratives can illuminate aspects of contemporary life and the human condition. In short, they can comment *upon* life, not just add *to* life.

Among many intellectuals, literary and art critics, this function of story is revered; works of art of all types are frequently evaluated on the basis of how well or how much they illuminate. Popular art, contrasted to fine art, is often thought to be inferior because it does not normally offer a novel perspective on life or ask probing questions. From this perspective, popular stories merely entertain or amuse, while truly great narratives make lasting contributions to our understanding of the human condition. Shakespeare explores human tragedy, for instance, while Bill Cosby merely provides laughs.

Although there is some truth to this generalization, it's also true that a given story's capacity for illuminating life depends somewhat on the social and cultural context of the audience, not just on the art itself. The amusement of one era can become the illumination of the next one—and vice versa.

If instructional narratives offer simple answers to questions, illuminative stories raise difficult questions: Why do people hurt others? Is justice really possible among humans? Can organized religion squelch true faith? Illuminative narratives try to reveal and clarify aspects of life and the human condition.

Christians can ask such questions just as productively as nonbelievers. Indeed, there is a valid role for such stories in the lives of Christians.

Unfortunately, illumination is overplayed in the secular world and undervalued among Christians.[47] Wolterstorff suggests that the cultural milieu of modern art has influenced even popular perceptions about the function of art. According to the modern artist, art should show us something new to "awaken (people) from their somnolence, or release them from their self-indulgent ideologies, or energize them into action."[48]

Since the art elites in society strongly support this function of narrative, they value stories that reveal something new about life and question people's beliefs. For them, the artist-storyteller is not so much an educator or instructor as a rebel or radical bent on shaking up the status quo—a kind of Old Testament prophet.

This has been true even on commercial television, where Stephen Bochco, a creative writer and producer, launched several series that questioned middle-class American life as much as they reaffirmed it. Among his successful programs were "Hill Street Blues," "L.A. Law" and "Doogie Howser." TV critics loved his programs because they seemed more "realistic" than others. But this realism was relative to typical TV fare, which sugarcoats life in the name of audience ratings.

Illumination and Economics

Why isn't there more illumination and less confirmation on television? The social institution of television determines whether the technology will be used to illuminate or confirm. In commercial television, the drive for audience ratings leads to stories that confirm (support the cultural status quo). Few Bochcos are able to convince the commercial networks to try new, illuminative programs. Nearly all commercial TV series are trying to resonate with audience taste, not question existing values and beliefs.

Fortunately, some television adaptations of literary works, even when lacking illumination, lead viewers to the novels or plays which they are based on. This role of the tube has been greatly undervalued.[49]

Public television offers a considerable amount of illuminatory drama, in both series and anthology shows. "American Playhouse" and "Masterpiece Theatre" are quite different but equally successful approaches to televisual illumination. "American Playhouse" seeks mostly native stories of American life, while "Masterpiece Theatre" uses more classical works with potentially universal themes.

On public television the stories are intended to reveal aspects of life, not merely to reflect or confirm life. In other words, public television has been significantly influenced by the "prophetic" ethos of modern art.

Christian Illumination

Unfortunately, in the arts today there is very little Christian illumination. Christian storytellers generally emphasize instruction and confirmation over illumination, as most religious art has over the ages. This is certainly true in Christian TV. Evangelicals are not always open to having their existing values or practices challenged. There is security in confirmation. This is exactly why televangelists are much more likely to tell tales of God's healing power than of people's suffering; upbeat stories attract optimistic Americans to become supporters of the ministry.[50]

The art world and entertainment industries need Christian artists who will both confirm and illuminate, but Christian audiences are weak-kneed about the latter. As in the wider culture, artistic illumination often leads to new perspectives on life, which in turn produce discomfort and provoke change. For many Christians change is hard to accept. Old-fashioned values and behaviors are more comfortable, and traditional ways of life are more secure.

We fail to recognize that what many Christians believe to be orthodox and trustworthy is often highly influenced by the winds of modern culture. It is easy, for example, to create a Christian soap opera by inserting a few God-fearing characters in the existing dramatic formula, as CBN did in the 1980s with "Another Life." It is much harder to create truly illuminative tales that probe the deeper meaning of the life of faith. Christian television producers have not done this on evangelical stations and networks. Yet it is precisely such probing that leads many readers to the writings of people such as C. S. Lewis and G. K. Chesterton.

As with the other functions of narrative, illumination is both God-given and humanly warped. Christians need to view stories that reveal evil and raise questions about it. But if they don't view discerningly, they are not taking seriously their responsibility as stewards of the creation. It's entirely possible for a story to illuminate sinfully and rebelliously, replacing God's privileged position as Creator and Lord with our own quest for personal glory and power. Modern artists and intellectuals often tend in that evil direction, in spite of all their claims to the contrary. Illuminative narratives are never neutral.

Conclusion
Christian viewers and producers of television often overlook the scope of God's cultural authority. Imagine a world where believers dedicated all their storytelling to the Creator. Where Christians realized that the tube was an important vehicle for helping to fulfill the mandate to be caretakers of the creation and guardians of the culture. Where evangelicals accepted and re-joiced in their role as shapers and consumers of godly narratives. Where television was more than the national jester.

The role of television as a storyteller should propel Christians toward accomplishing these goals in four ways.[51] First, Christians should break the bad habit of using television only for amusement or instruction. Like the nonbelievers who surround them, Christians are apt to seek and expect little more than diversion from the tales on the tube. And when they do seek more, it is usually moralistic instruction, which characterizes the vast majority of evangelical film and video productions.

Christians should look for programs that help them critically understand the world in which they live and that validate the historic Christian faith. I find it disappointing that few evangelicals watch serious drama on public television, cable channels and videotapes. The Christian viewer can find a richer, more meaningful life in the balance among amusement, instruction, confirmation and illumination.

Second, Christian viewers must become far more discerning about how stories influence their lives and shape their faith. Too many Christians wrongly believe that the tube is merely entertainment for mindless diversion. Every story is not just a diversion from the world, but also a door to another world. Stories are part of the language of the culture, and the tube is a popular voice.

Discerning viewers will ask themselves how programs fit into their daily lives and whether the stories' lessons or values are in harmony with the Christian faith. It is naive to expect all shows to preach the gospel, but it is similarly naive to assume that television never challenges Christian values and beliefs.

Third, the Christian community needs to cultivate its own local storytelling in the face of television's enormous popularity. We cannot be only consumers of others' tales; we must create and pass along our own stories in families, churches, schools and neighborhoods. Not all of us are called to make television programs, but each of us can participate in home-grown narratives that reflect the Christian faith, teach its implications and even amuse. In fact, telling our own tales helps us understand how stories work and facilitates sharing our lives more fully with the people we love. When we lose track of our own story, we are most ripe for others' propaganda.

It is a sad fact that when we watch a lot of television, we tend to neglect our local responsibilities. The tube steers us outward, beyond our communities and families, and into a prefabricated world of consumer tales. A study conducted by Louis Harris and Associates in 1987 found that VCR ownership had skyrocketed 234 per cent in the previous three years, while attendance at dance performances, theater, classical and pop concerts, and opera and musical theater was down between 14 and 38 per cent.[52]

If we're not careful, we will all lose track of our own stories: who we are, where we came from, how our parents reared us, what we truly believe and how we shall then live. My point here is that our own family, church and community histories are essential stories for maintaining our faith across the generations. So are ethnic and religious stories. We should not trade these home-grown parables entirely for those produced by the entertainment industry.

Fourth, Christians need to be far more supportive of gifted storytellers who seek to please God and serve humankind. There is a significant dearth of Christian tellers of tales across the various media, from the printed word to cinema and television. Too often such gifted people are driven from communion with the faithful because of the narrow-minded attitudes of the Christian community.

Christian writers and producers for television need our encouragement, and sometimes even our financial support. It is simply not possible to have a major impact on American culture without cultivating Christian bards. Television, as the most popular storyteller of our day, needs gifted Christian storytellers to amuse, confirm, instruct and illuminate—even to tell stories to audiences of penny-tossing adolescents.

I learned this the hard way in my own teaching and writing. There is something profoundly significant about narratives—something that allows humans to experience life more fully and richly than animals do. Through stories we recall our past, learn from it, and plan our future.

To be human is to be a storyteller. Clearly God made us this way. We should enjoy and celebrate storytelling, even on television.

3
GRAZING
VIDEOTS/
Why We Only "Watch" TV

Dear children, keep yourselves from idols. *(1 JN 5:21)*

IN 1991 THE CITY OF ST. LOUIS, MISSOURI, CAME UP WITH AN UNUSUAL solution to disciplinary problems on public school buses: the nation's first "video school bus," with five TV monitors, eight built-in speakers and a VCR.

Apparently the video bus worked. According to the bus driver, younger children remained glued to the 10-inch screens as if hypnotized. Said the driver, "When I get to school, they want to sit on the bus and continue to watch." And on the return trip all the students wanted to be dropped off last.[1]

Although the school system justified the video bus partly by showing educational tapes, the success of the project raises an interesting question: Why would TV sets tranquilize these inner-city kids? Is the tube an opiate? Or do we merely learn to view it passively? Does the *technology* itself cause such passive watching?

The technology itself doesn't bear full blame. The *way* we watch television, *where* we view it and *why* we watch all shape its impact on our lives. Television cannot be redeemed until viewers accept their responsibility to use the medium wisely.

Unfortunately, home viewing patterns and newer video technologies encourage mindless viewing as well as undiscerning judgments about what, when and how much to watch. Like the proverbial fish in the water, viewers are unaware of the extent and significance of the television environment in which they live. To a greater or lesser extent, we are all like those students on the video bus, passively soaking up programs.

Twenty years ago virtually no college students brought television sets to campus. In fact, some schools forbade it, partly for academic reasons and partly for a very practical one—the electrical circuits in dormitories couldn't provide adequate power. But a younger generation of college students, raised on "Sesame Street," sees the tube as part of its lifestyle.

A survey conducted in 1991 determined that about three-quarters of undergraduates had their own televisions at college. Half of them had a VCR as well, and half received cable. They watched an average of eighteen hours of television weekly. While their professors may have asked them to think, the tube, as a social institution, had its own message: Watch. As Bruce Edwards, Jr., put it, television seems to offer only this one commandment: *Watch.* "Watch and do not speak. Watch and do not act. Watch and remain still, placid, passive. Watch: unself-consciously, docilely, imperturbably. Watch: losing yourself in the mediated world, ignoring your own."[2]

Television seems to have invaded every area of life, from school buses to college dorms, from bedrooms to waiting rooms, airports and campgrounds. And along with technology the audiences import into these places a way of using the set. They learn to "watch TV" instead of viewing it critically. As Solomon Simonson put it, "*Passive lookers* represent the uncritical majority of the audience."[3]

Audiences learn to use the set to pass the time rather than to stimulate thought, elicit conversation or explore emotions. Television watching has become the major leisure activity in the United States, representing over fifteen hours per week of "free time" for adults—five times more than the next most common leisure activity, traveling.[4] Studies suggest that in November, when the new shows are being extensively promoted, adults watch about four hours and forty minutes daily.[5]

Scholar Margaret Morse calls television viewing a "distraction," like a trip through a shopping mall or a ride down a highway.[5] Television critic Michael Arlen compares watching television with flying in a modern airplane. Both are a "passive," "experienceless voyage."[6]

These metaphors are apt for much television watching today. For most people, watching the tube is like sitting in a jumbo jet and staring mindlessly out a window. The changing scenes are a diversion from real life, which begins anew when the voyager "deplanes."

Sometimes I prefer to compare the way many individuals look at the tube with the way a cow watches passing cars. The eyes are open, and some type of sensation is generated in the brain of the bovine. But there is no critical mental activity, no power of interpretation or evaluation.

Our viewing patterns and habits, as part of the social institution, contribute to the passive use of TV. For example, we know that nearly all households have a set—some have as many as five. Most of them are located in living or family rooms where members congregate for leisure activities. A bedroom is the second most likely room for a television, followed by kitchens and other rooms. Most adults and children watch every day. Viewing is private and informal, not a public occasion.

The tube has been almost thoroughly integrated into the routines of domestic activity; the set is turned on before school, and many people sleepily watch late-night news shows. Some glance at the set while folding the wash; some use audio for company in a quiet house during a lonely day.

In short, we have adopted lazy, unreflective television habits. Instead of viewing the tube, we watch it. Christians will never redeem their viewing habits unless they engage their minds and hearts. The social institution of viewing has to be changed before the technology can begin to catch up with its potential.

Privacy

The privacy of television viewing significantly influences what programs we watch, and how we watch. If television were watched publicly, we would be far more concerned about other people's perceptions of our viewing habits. Movie selection, for example, has long been affected by what kinds of movies individuals think are socially appropriate. The fact that filmgoers had to "go out" to see a movie affected their decisions about which films to see. Public viewing increased patrons' sense of social propriety and personal conscience. No one wanted to be seen entering a theater that was playing a morally questionable movie. Conscience was strongly shaped by social taboos.

Thanks to the relative privacy of television, adults are now likely to choose their television programs whimsically and thoughtlessly. Cable and the VCR do not require private film viewing, but they encourage it. Even some evangelicals will use cable or the VCR to watch the same R-rated films that they would not want to be seen viewing at the local movie theater.

Private television viewing reduces the likelihood that community or societal norms will be used by individuals in their decisions about what to watch. Thus, TV viewers are less likely to think about what to watch and how to watch it.

Privacy is not all bad. Sometimes social standards, especially within a subculture like the evangelical community, place taboos on excellent programs.

Of course, the peer group's influence on teen viewing remains strong. Videotape rental, for instance, is still heavily guided by peer pressure among adolescents, who want to demonstrate to their friends that they have seen particular movies.

Adults are less vulnerable to this type of peer pressure and significantly more individualistic in their viewing habits. Although the publicness of films will always influence what many adults select, private television viewing guarantees far more personal freedom in choosing programs and films. Without a strong sense of responsibility, this is a dangerous trend.

More important, privacy tends to destroy the communal experience of viewing. Stage and film traditionally relied extensively on the existence of an actual audience, not simply on a distant collection of individual viewers. As television personalized and individualized viewing, the nature of the audience changed. Domestic viewing largely eliminated interaction and social dynamics among members of the audience. In one sense this was good, for individual viewers were no longer distracted or offended by the antics of other people in the audience. In another sense, however, individuals lost the social cues that taught them how other people responded to particular scenes and specific dialog. The television viewer was left to figure out what was happening on the tube and respond to it alone.

Aware of this problem, many television producers decided to add fake laughter to the soundtracks of their programs, simulating the expected response of a real audience. Canned laughter was meant not only to give viewers the impression that they were viewing the show with other people but also to instruct them about when to laugh. Most viewers do not laugh simply because they hear artificial laughter in the program's audio track, but they do get a sense of how funny a program supposedly is for other people and how humorous it should be for them.

Anyone who has attended the studio taping of a situation comedy knows what I mean. Such public viewing is an entirely different experience from watching the show at home—and not just because of the excitement of seeing a show live. Normally at these tapings someone entertains the audience between scenes and retakes. The studio audience is exhorted to enjoy the program, reminded how humorous the show is and prompted to laugh at particular lines, even when the actors have already bungled the dialog numerous times. In short, the studio audience is manipulated in order to orchestrate the private responses of viewers at home.

Invited into American homes, the tube fostered private viewing while it brought domestic disharmony. Americans argued about where to put the set and what to watch. Televisions interfered with the ideal of what a suburban home's interior should look like, yet simultaneously they became a symbol of middle-class prosperity.

By 1990 about 43 per cent of American households still put the most-watched set in the living room (perhaps more appropriately called the "viewing room," since furniture was typically organized for easy television viewing, not for enhancing family activities), but over 34 per cent of homes had one in a bedroom as well. Because the tube has become such an object of personal attention, it never has really made peace with either home design or family life.[7] The set is always there, attracting attention and diverting individuals from family activities. Surveys have shown repeatedly that each day Americans most look forward to spending time with friends, neighbors and co-workers, but they actually spend far more time watching the tube.[8]

More than other forms of drama, television programs privatize viewing. The medium creates the impression that viewers are part of public life because millions of others are watching the same program. Actually, though, television's privacy makes viewing private and personal. A movie or play will typically be seen in conjunction with dinner or another social event; almost none of this takes place with the tube.

There are exceptions: major sports events, for example, bring people together at the same home for viewing. Overall, however, television is viewed and evaluated privately and personally—if it's evaluated at all. In these senses television viewing is more like reading than like attending dramatic performances.

Technological developments have further encouraged private viewing. Large-screen sets offer the potential for groups of viewers, but in the typical family room such sets overwhelm the senses, making it unlikely that viewers will communicate among themselves about the programs. Also, there are typically few opportunities for communal viewing, because families and friends are too busy with their own activities. The newest micro-screen sets inherently individualize viewing, because only one person can comfortably watch the set at a time. Paradoxically, these small sets can be used most easily in public, especially at sports events; they turn public viewing into a private activity.[9]

If television had become another publicly viewed medium it would be very different today. In the late 1940s, neighborhood taverns lured local patrons by installing sets above the bars.[10] For a short while, in urban ethnic communities television was indeed part of public life. More surprisingly, it was even part of family life, as entire families went to pubs to see their favorite shows. The shows of Bishop Fulton J. Sheen and comedians Milton Berle and Jackie Gleason, along with professional wrestling, were popular as friends congregated for the broadcasts.

But as television sets became more affordable, and personal incomes increased, Americans purchased sets and installed them in their own homes. Even in urban ethnic neighborhoods, viewing shifted from public to private locations.

Casual Viewing

Partly as a result of the privacy of viewing, television viewing is less a special occasion and increasingly part of the daily flow of home life. The set is always "there"—available whenever someone feels like viewing it.

By contrast, going to a play or movie is still a much more formal occasion that requires special plans and premeditated actions: when to go, what to see, who to go with, which babysitter to get and so on. Broadcast television, like cable and the VCR, has virtually eclipsed forethought and formality. Viewing the set is just another domestic diversion, not an activity to be taken seriously. Of course, there are a few exceptions to this rule, such as viewing Super Bowl broadcasts, televised premieres of major Hollywood films and the concluding episodes of some long-running series. Overall, however, watching the tube is more like raiding the refrigerator than preparing a balanced, nutritious meal.

By scheduling shows nearly around the clock, television stations and networks encourage private, informal viewing. Whenever a viewer decides to watch television, he can be sure to find something on one of the channels used in his area. Only in smaller cities do stations regularly shut down the transmitters overnight, and even then viewers can turn to tapes rented from local shops or video recordings of earlier programs—if not to an array of cable channels that never stop programming.

Children watch for ten minutes while Mom or Dad finishes fixing dinner. Teenagers see fifteen minutes of a show while waiting for peers to arrive. Newscasts and sports programs are tuned in for only a portion of their full length. In these ways viewers casually make television a part of their daily lives—like picking up a coffee-table book at a friend's house or *People* magazine at the doctor's office.

Domestic Distractions
The implicit domestic rules for viewing suggest to us that television does not deserve special attention. Audiences approach the tube with little intellectual and emotional preparation or expectation; distractions are part of watching. Cable channels, such as Home Box Office, which runs primarily films and music concerts, have improved things somewhat, but watching television is far less focused an activity than going out to attend a play or see a motion picture.

One study found that 70 per cent of viewers watch television primarily as "a way to pass the time," and only 26 per cent planned their viewing in advance. Moreover, 40 per cent said they left the room repeatedly during programs. Forty per cent also said that they give programs "some" or "hardly any" attention. About one-fifth of the people who have the TV set on are not really watching it, although the percentage changes considerably with various programs.[11]

Sports "replays" on television are provided partly to give viewers a chance to see what they missed because of inattention; few replays are actually needed for making judgments about the calls of game officials. In short, we "watch TV" instead of attentively viewing specific programs.

Viewers are not captive audiences, but easily distracted individuals whose concentration can quickly be interrupted by other people and by the need to attend to other duties or activities. The commercial television industry therefore devises strategies to keep viewers both in front of the set and tuned in to particular programs. Attracting audiences is much more important than encouraging critical viewers. As we shall see later, the networks invest considerable energy in determining how to schedule their programs during the evening and throughout the week in order to maximize this "audience flow."

From a social standpoint, the problem of keeping people attentive to *any* television program is very significant. There are telephone calls, doorbells, questions from spouses, crying children, meal preparations and hundreds of other legitimate interruptions. The set is positioned in most homes for easy viewing, and it is frequently turned on, but the programs easily become background noise and meaningless flickering images. Television becomes domestic ambiance instead of a communications medium or dramatic form.

As producer Sheldon Leonard has suggested, television does not

command the depth of people's attention. One who goes to a Broadway show surrenders a part of his individuality and becomes a component in this thing called an "audience." He becomes part of another entity, and until the show is over and he has to re-orient himself to the reality around him again, he remains part of something else that is not real. The audience does not relinquish this sense of individuality in the home. A phone will ring, a child cries, and they are constantly next to the reality of their own lives as they watch the set. They cannot immerse themselves totally in what they see.[12]

Many viewers today darken the viewing room in order to minimize visual distractions and focus attention on the set, but it hardly helps. The distractions are still there, and our viewing attitudes remain the same—TV is just entertaining diversion. In the first few decades of television this was generally not the case.

The rapid proliferation of VCRs in the late 1980s transferred some of the ambiance of the movie house to the family room. Growing numbers of families assumed that rented videotapes should be viewed in relative darkness, even though television sets actually produced images of equivalent quality in most lighting conditions. Viewers began simulating the atmosphere of movie theaters, aided in a few homes by the new large-screen sets and stereo sound. Yet viewers' mindset actually changed little, however, because of the normal interruptions and informality of domestic viewing. In fact, the carry-over from home viewing to the film theater may have had more impact than the reverse influence. During the same decade, growing numbers of theater patrons complained about patrons who talked during films.

Commercial Interruptions

Commercials also contribute to casual, inattentive viewing. Repeated program interruptions suggest to viewers that shows are unworthy of concentration and extended mental focus. Moreover, the lighthearted, silly tone of most advertisements carries over to the programs, minimizing the impact of even some of the most somber television drama. Late-evening films on television are chopped up by commercials for "classic country-and-western hits," while prime time is minced by a hodgepodge of advertisements for over-the-counter drugs, fast-food restaurants and diet colas. Images flow on the screen without stop, but rarely is a program free from the distracting interruptions of the sponsors, except on pay cable and public TV channels.

Television drama has been greatly affected by commercial interruptions. In the early years, commercials were run only once during a half-hour broadcast, and a number of shows escaped commercial interruptions altogether because advertisements were placed immediately before and after the program. As the networks shifted from single-program sponsors to multiple "spot" advertisements, they increased the total number of commercials per hour. Eventually, advertisers' complaints that too many commercials were being run back to back led networks to increase the number of breaks in each program—two or three for half-hour shows, and five or more for hour-long programs.[13]

Today producers write scripts to keep viewers tuned to the channel during commercials. The first eight minutes or so of a sitcom are designed to create enough confusion and misunderstanding among characters to hold viewer interest through the first commercial break. Game shows devise formats that fit the break periods as well as the half-hour program length. News broadcasts often delay one or more important stories until after the first round of advertisements, teasing viewers with a visual or spoken reminder about the "upcoming story." And made-for-television movies have forced themselves into "an unnatural cliffhanger formula in which climaxes are contrived for every fifteen minutes or so in order to hold viewers through a commercial break."[14]

Sports shows are one of the few types of programs whose content dictates the placement of commercial breaks, rather than vice versa. Yet game starting times and the length of pauses in the action are now frequently altered by the networks, especially during championship series. Often the extra "time outs" for commercials are disconcerting for fans in attendance at a game that's being broadcast.

All in all, the chaotic structure of television content, from commercials to shows and back again, mirrors the flow of contemporary life. With so many "activities," it seems that many people are barely able to take any of them seriously, even "breaks" for relationships. The distractions to television viewing mirror the real fabric of life.

Beating the Clock

The predictable lengths of television programs probably inhibit viewer attention also. Half-hour and one-hour programs are standard fare. Only feature films and sports events regularly run odd lengths, in spite of networks' attempts to make them conform to scheduled half-hour (really twenty-two-minute) blocks.

From the perspective of the commercial broadcast industry, programs are like pieces of a jigsaw puzzle that must be fit together effectively to build and hold audiences through the day. Network programs begin on the hour or half-hour to attract viewers from other networks, whose shows are ending at the same time. Canceled programs may be syndicated as reruns, which easily fit into open half-hour and sixty-minute slots.

As a result, television producers put together shows of standard lengths regardless of the amount of time actually needed to tell a story well. Fifteen minutes of good sitcom material will be expanded to a half-hour show by the addition of several superfluous scenes and a few extra gags. Important aspects of a film's plot may be edited out so that it can be run in a two-hour slot.

Television producers Richard Levinson and William Link once told of adding thirty minutes of material, some of it extraneous, to their award-winning teleplay *The Execution of Private Slovick,* simply to please the network. They argued that a motion picture can usually be cut to its best length, but a television film has no elasticity because networks provide detailed requirements for footage and commercial breaks. "This is one of the reasons why television shows, even the segments of a series, may seem long or excessively sluggish," they conclude. "To get a variance is almost as difficult as securing a papal dispensation."[15] Television programs are written for the clock, not solely for the audience or the story itself. The action generally occurs at predictable intervals, offering viewers few surprises and demanding minimal attention.

In a sitcom, for example, the characters and setting are introduced and then the comic situation is created—usually through a misunderstanding or confusion among characters. Finally, the complications caused by the misunderstanding are resolved. Admittedly, this formula is sometimes used in novel ways and developed with a sensitivity to character. But the approximate time at which each of these events must occur in the plot is held relatively constant by the placement of commercial breaks during the twenty-two minutes of actual program time.

The television industry's demands for standardized program periods and a specified number of commercial breaks guarantee that the plot's timing will be kept relatively constant and that innovative uses of the formula will be rare. The unfolding of events is relatively predictable, providing viewers with an easy entry into the middle of shows and an easy route out of boring ones.

In recent years the clock has become increasingly unimportant to viewers. Instead of tuning in at particular times to see particular programs, growing numbers of viewers watch the tube only when it's convenient or when they have time to waste. As odd as it seems, fewer and fewer viewers care whether or not they see the beginning or ending of a program. Aided by remote-control channel selection, viewers increasingly bounce from one channel to the next in hopes of finding something to relieve their boredom at that moment.

Thus viewer expectations have shifted—away from well-crafted narratives and toward exciting, or at least visually interesting, scenes. Indeed, many television dramas today are little more than a series of attention-getting scenes without any overall theme. As columnist Ellen Goodman has written, "The only thing that television itself asks is that people watch. There is something intrinsically passive about this."[16]

Money over Mind

The low cost of television viewing inhibits attentive, reflective viewing as well. Like free food at a smorgasbord restaurant, the platter of television shows allows viewers to indulge their gluttony. Simply put, television is the cheapest form of professionally produced drama available on a mass scale. Why should viewers take it seriously, when there is so much of it and it costs them so little? Even the expense of a new set is rarely considered in relation to its use for an individual program or series. Set sales increase before some attractions, such as the World Series or the Super Bowl, but these periods are exceptions. For most people the important thing is to have a set, not to have particular programs.

Few viewers seriously consider the financial cost of viewing a show, or even the cost of watching an hour or two of programming—as they might consider the expense of going to a movie or the theater. If costs for individual programs were listed, like à la carte items on a menu, viewing would likely fall precipitously.

Imagine if every set had a slot for dollar bills. How many dollars would the typical viewer be willing to put in the slot for "Family Matters" or "Full House"? For the news? Sports? Without such a clear sense of the relative cost of television viewing, viewers simply do not compare the merits of shows, let alone the relative merits of watching the tube versus other activities.

People like a free lunch, and television seemingly provides it by not associating any direct costs with the selection of particular programs. So viewers do not highly value or appreciate what they watch on the tube. This is precisely why pay-per-view television, which charges cable viewers for specific programs, has not been enormously successful. Only blockbuster events, especially big-name rock music concerts and prize fights, are very successful pay-per-view programs.

Cable television, now in almost two-thirds of American homes, encourages largely the same attitude toward the cost of viewing. Cable generally disassociates the monthly service charge from the cost of particular programs. Viewers sometimes purchase cable initially in order to receive programs not available on broadcast television, but soon the monthly rate is only vaguely related to actual viewing. After all, the cable viewer pays the same amount no matter how much he or she watches.

The only incentive is to watch more television to justify the expense—so people do watch more television after they get cable installed. Yet the additional viewing that takes place in cable homes is not necessarily more attentive or more enjoyable. Once the novelty of cable television wears off, viewers simply approach it as they would the old-fashioned, over-the-air programming—except that they watch more of it.

Interestingly enough, video rentals are expensive enough to have an effect on viewing. Two or three dollars for renting a videotaped film will not make most middle-class viewers feel that they should carefully dedicate their time and attention to viewing, but it is enough to help make selection a bit more discriminating and video viewing slightly more serious than traditional television watching. Films watched on the VCR are probably viewed more attentively than traditional television programs are, largely because acquiring and playing tapes is more costly and purposeful than watching episodic television series or other prime-time fare.

Lazy Looking

Among all mass media, television is one of the easiest to use. This, too, encourages informal, inattentive, mindless viewing. "Watching TV" requires little more than keeping one's eyes open and focused on the set. Compared with reading, which requires one to hold the book and follow the words with one's eyes, television viewing requires little physical effort.

Mentally, too, reading is more taxing than viewing the tube. It probably even takes more mental effort to listen intently to radio drama than it does to watch television. At least radio plays expect far more of the audience's imagination than does television drama. The viewer need only see some of the images and understand some of the dialog in order to follow the typical TV story. Simply put, television, more than most media, facilitates mindless watching both because of the nature of the technology and the way viewers learn to use it.

Effort, like cost, relates to perceived value. Suppose viewers approached the tube with the intensity of an evening at the opera or theater. That's very hard to imagine—yet the technology itself hardly requires viewers not to take the medium seriously. But an enormously influential *institution* has developed around the technology to define its normal uses and purposes. This institution does not want the tube to be taken too seriously, except as a money-making enterprise. In commercial broadcasting, where the drive for audience ratings and advertising revenues is taken very seriously indeed, the programming is merely a trivial means to very significant economic ends. The greatest effort is put into figuring out which programs will attract the largest audiences, not which shows will be viewed intently by audiences. In public broadcasting, by contrast, much of the adult-oriented programming is viewed attentively, is produced by people who take the artistic side of their craft very seriously, and is even reviewed more carefully by critics.

The effort people put into an activity undoubtedly affects their attitudes toward that activity, and television is no exception. When the tube is defined as mere amusement, few viewers try to understand or evaluate the content of programming. If television is only a neutral way of passing time, then it surely should not be taken too seriously.

But if viewers are willing to put some effort into viewing television, they will expect more from the networks, stations and cable companies, even if they watch less frequently. It stands to reason, then, that overcoming lethargy is also part of the goal of redeeming television.

Unpromising Technologies

In spite of all the public hoopla in recent years, the new television technologies have usually dug the audience deeper into the same uncritical viewing patterns. Even with the innovations of satellites, cable, VCRs and remote-control channel selectors, program producers and viewers have maintained the attitude that television has little to offer beyond private, informal, free and effortless viewing of mere entertainment. With each new technology, viewers will generally watch more. For example, households with pay cable channels have the set on more than those with only basic cable, which watch less than households without cable.[17]

But greater viewing does not mean more discriminating program selection. The latest technologies, including projection and multi-image sets (picture-in-a-picture) with stereo sound, hold out little hope for reforming the medium unless producers and users share a new vision of what television could be as a source of drama and information.

Satellites and cable have expanded the variety of programming, but have not redeemed television viewing. These technologies offer quicker news reports, but not necessarily more accurate, comprehensive or even meaningful ones. This was obvious in early 1991, when the three major broadcast networks and especially Cable News Network (CNN) aired instantaneous reports of Iraqi missile attacks in Israel and Saudi Arabia. Yet the networks failed miserably to put the Gulf War into a reasonable historical, political and economic context. They raced to get the latest missile information, not to explain what the war meant to people around the world.

This was a potentially very dangerous situation, because CNN and its competitors became an international nervous system sending messages instantaneously throughout the world. Erroneous reports can have immediate effects as governments react to what they see on the tube, not necessarily what is really happening. A military consultant to CNN, Major General Perry M. Smith, told how during the Gulf War he was asked to make live interpretations of videotape from the front that he hadn't even had a chance to preview. "Throughout the entire war," said Perry, "I never once slowed down the process in the name of more thoughtful analysis."[18] In other words, even the makers of the programming did not think in advance about what they were broadcasting.

Second, cable and satellites have greatly expanded the effortless home viewing of feature films, but they have made no significant distinctions about the moral, artistic or spiritual quality of those movies. More channels required more movies, and it seemed that any films would do the trick as long as a few especially popular movies were thrown in now and then to keep subscribers happy. Channels such as HBO, Showtime and Cinemax all played this game. There are now so many movies on cable TV that it is virtually impossible to select and view them wisely. We have quantity, but viewers' ability to discern quality has not been enhanced.

There is no doubt that variety will increase. In December 1991 a young married couple in New York State became the first cable subscribers in the world to get 150 channels, including 57 channels reserved for pay-per-view movies around the clock. The system had so many program options that it was unrealistic to expect the cable company to print a program guide, which would look like a telephone book. Moreover, the advanced cable system permitted viewers to watch two pay-per-view movies at the same time—so individual viewers could flip back and forth or families could view different movies simultaneously.

The wife of this cable-TV family exulted in the complexity of it all, including the four remote-control devices: "I don't know what to watch. . . . It's better than going to a video store. It's, like, you don't know what to watch. You can catch a movie every 15 minutes."[19]

Discriminating viewers can find many excellent films on videotape and cable channels, but the scope of the technological developments has made careful and studied viewing choices unlikely. Who has the time to become adequately informed about all programming options? Isn't it easier to flip around for a show that appears more interesting or entertaining than the other ones?

A former head of programming at NBC, Grant Tinker, said in 1990, "There is too much [television]. Maybe that is part of what's wrong with it. . . . There are so many viewing alternatives. When I was a kid and television arrived, it was magic."[20] In the United States that attitude seems old-fashioned, even undemocratic. But actually, more options do not guarantee better programming; they do make it harder for viewers to make informed decisions.

More than these technologies, however, the simple remote-control channel selector has fueled lazy, ill-informed viewing. In one sense this type of technology is the ideal mechanism for maximizing pleasure for home viewers. Since viewing is invariably interrupted by extraneous noises, unexpected phone calls, needy children and the like, the remote control enables the individual viewer to adjust effortlessly to distractions. If one loses one's place in a story, one can quickly find another to pick up on. If the phone rings, one can hit the "mute" button. Similarly, if commercials are uninteresting the viewer can skip to another channel and return later.

In the face of all the choices available, it's easier to use the remote control to hunt for entertaining shows than to plan viewing. This technology opens up the smorgasbord with little or no demand upon viewers to select carefully the shows they watch, let alone the ones they don't watch. Remote controls accept viewers as they are and then help them to be even more that way—lazy and indiscriminate.

Film director Steven Spielberg laments the impact of television-style watching on the motion-picture audience. Claiming that the "new generation learned movies from TV," he argues that "MTV has been one of the most destructive influences on the current crop of movies. . . . And the news has had a negative effect, since we're getting stories of epic disasters in sound bites no longer than a couple of minutes. So people are flipping channels as quickly as the news is changing stories from joyful to tragic, as quickly as MTV is giving you a four-minute video on a song and bombarding you with eighteen-frame cuts."[21] If Spielberg is right, technology and social institution are turning the movies into television programs.

Grazing Videots vs. Redeeming Image-Bearers

The image of today's television viewer is appropriately captured by the term "grazing videot." Like cattle out to pasture, viewers lazily sit in the comfort of their reclining chairs and tap the selector switches on their remotes, sampling a drama here, a comedy there, a promotion, and perhaps a scene or two of a documentary. The word *videot* comes from novelist Jerzy Kosinski, who warned in *Being There* that television was producing passive people who only react to life rather than act upon it.[22] Grazing videots are viewers with a greater passion to satiate their televisual thirst than to shape the medium in ways that would advance culture or improve society. In other words, grazing videocy distorts the creational potential of television.

Mark Crispin Miller believes that the "illusion" of real viewing choices is crucial to understanding the mindset of audiences. "Everybody watches it, but no one really likes it," says Miller. "This is the open secret of TV today." We share an illusion about the real benefits of contemporary programming. The tube promises choice but "shows us nothing but the laughable reflection of our own unhappy faces. . . . It celebrates unending 'choice' while trying to keep a jeering audience all strung out."[23]

Surely Miller exaggerates his case against "all" television. Nevertheless, his words ring largely true in the age of grazing and remote-control channel selectors. A "choice" of nearly identical fare, selected thoughtlessly, is hardly a choice at all. It is clearly not artistic or cultural freedom.

The point here is not to condemn amusement or to call for intellectually elitist programming. Rather, I wish to argue that television *viewing* holds far more potential for redeeming television than does television *watching*. Watchers are mere grazers; their enjoyment is primarily in the act of watching the set, not in the act of viewing particular programs.

Christians are not called to "watch TV," but to make and view programs that are morally, artistically and spiritually superior. Viewing requires discriminating decisions based on more than the momentary pleasure derived from using the remote-control selector. It means calling on the mind and the heart to redeem the medium.

The touch of the Creator is apparent through our culture and society when Christians live redeemed lives. For members of a television world, this means casting off the selfishness and laziness generated by the Fall and redeeming their lives in service to the kingdom of God.

No mass medium has ever been such a pervasive temptation to mindlessness and indiscriminate actions as television. Televisual grazing is not only a sign of laziness but, more important, a reflection of our failure to bring all things under the lordship of Christ.

Conclusion
In order to become more than grazing videots, Christians must change their attitudes toward watching television. There are four things we can all do to encourage viewing and discourage videocy.

First, Christians must do their part to shift television viewing from a predominantly private, individual activity to a public and communal one. The tube cannot possibly be redeemed as long as it is reduced to an essentially private and personal phenomenon.

The Christian community needs to make television part of its public discourse and discussion. As things now stand, few people know or care what others watch, and television is rarely considered in Christians' dialog about the broader culture and society. (News and sports are exceptions.) Television drama enters Christians' public life normally only after a particular show or event has stirred up considerable controversy. This happened with the televangelism scandals in the late 1980s. It also occurred with the salvos fired at Fox's "Married . . . with Children" and "The Simpsons" a few years later. In both cases the lack of a persuasive, articulate and informed Christian assessment of the Fox series led to little more than additional publicity for the programs. As they often say in Hollywood, bad publicity is usually better than no publicity.

I would encourage Christians, in families, churches and fellowship groups, to begin discussing television. We need to share our ideas about particular programs, our strategies for making judgments about them and our methods for finding the best fare on the tube. Along the way we will naturally be shifting viewing from the realm of personal choice to the far better level of community critique and collective action.

This will likely mean that we watch less television, since there must be time for discussion and even prior reflection. But it will also lead to more communal viewing: going to friends' homes to watch particular shows or trading tapes with them. Certainly not all, or perhaps even most, of our viewing will be done publicly with others, but we can still make television a much more significant part of the public life of the church.

Second, the Christian viewing public needs to create far more formality around the use of the tube in the home. Most viewing is now so extremely informal that it readily fits comedian Fred Allen's phrase: "chewing gum for the eyes." If that's all we think of the tube, then we really ought to get rid of it—for television is actually values and beliefs, ideas and actions, purposes and motivations, not just chewing gum.

But the Christian community can lead the way in placing television in a more formal viewing context. Even amusement should be taken seriously. Instead of using the tube to fill in empty time, or watching bits and pieces of shows, we can gather around the set for special programs that merit our time. We can teach our children not to use television lazily, as an impromptu babysitter or daily companion. They should be taught to plan their viewing in advance, making hard choices about which programs are worth their time. Once television viewing is turned into a more public and formal activity, even a special occasion, we will be ready to overcome mindlessness in our use of the tube.

Third, Christians need to apply more discernment in selecting and viewing programs. A good rule of thumb is to balance viewing among the various types of storytelling—amusement, illumination, instruction and confirmation. It takes thoughtful effort to decide what is worth watching when the programs are changed regularly and the number of viewing options is increasing every year. Viewing can be fun, but deciding what to watch usually requires some work. So does critical viewing. Grazing is not the way to select and view programs.

Fourth, Christians should accept new television technologies only after developing a strategy for using them appropriately. It's too easy to assume that a family will figure out how best to use cable or a VCR *after* getting it. In such situations the lack of forethought almost always leads to problems; viewing patterns are established before the family has time to think about the issues involved.

If we are going to invite these technologies into our homes, we need to be clear in advance about the benefits and pitfalls. Even the decision about how many sets to have is crucial to redeeming a family's viewing.

As a general rule, every new technology expands the potential for good *and* evil. In other words, technological improvements almost invariably increase the difficulty of making decisions about their use. The remote-control channel selector, for example, makes it easier to eliminate unwanted shows—but also offers an effortless way of tuning in evil or offensive fare. Similarly, the VCR opens up to any adult viewer an incredible spectrum of tapes available for purchase and rental.

Part of the responsibility for redeeming television lies with viewers. Much blame surely lies with producers, but it would be unfair to blast the networks, stations and production companies without asking viewers to examine the logs in their own eyes. Evangelicals like to criticize the entertainment industry, but there is abundant evidence that the business could not survive without the uncritical viewing standards and practices of the general public.

My own interviews with hundreds of evangelical families suggest that Christians and the general public watch essentially the same kinds and amounts of television programming. The only exception is explicitly religious programming, especially Sunday-morning fare, which is viewed overwhelmingly by evangelicals. Ironically, these programs appeal to viewers for funds in the name of reaching unbelievers.

In May 1990 there was a rather telling example of the need for redeeming viewing. In a joint effort by local Channel 5 and the Cincinnati Reds baseball team, the station on a Friday night ran an otherwise blank screen for thirty minutes with a message: "Take this time to read together." Believe it or not, seven per cent of the city's residents with their sets on at that time were tuned to the "blank" program—nearly four per cent of all television homes in Cincinnati.

Perhaps viewers actually were reading while the message was posted on their sets. More realistically, it would appear that viewers did not care what was on. The programless show beat both the "MacNeil-Lehrer Report" and a rerun of "The A-Team" that were on other channels at the same time.[24]

As things now stand, millions of viewers are mindless grazers in search of images that can satisfy only briefly. Home alone, armed with their remote controls, they indeed approach television as if it were a fast-food smorgasbord—a free ride on the video bus.

Christians need not take part in this type of viewing. By using television thoughtfully, they can harmonize the Cultural Mandate with the most popular storyteller.

4

TELEVISUAL LITERACY/

How to View Persona

The eye is the lamp of the body. If your eyes are good, your whole body will be full of light. But if your eyes are bad, your whole body will be full of darkness. (MT 6:22-23)

IN 1987 COLONEL OLIVER NORTH APPEARED ON NETWORK TELEVISION BEFORE a congressional committee investigating the Iran-*contra* scandal. There were allegations that North and others had been involved in selling arms to Iran illegally in order to help free American hostages in the Middle East as well as to finance secretly the right-wing *contra* fighters in Nicaragua.

As Bill Moyers described it, when North took his place at the hearings "his people insisted that the network cameras be a little below him so that you'd see this jutting Marine's jaw looking up with that bold and courageous defiance at this tribunal of elders. . . . He made that theater by the angle of the camera." Soon afterward, American and Canadian students in the Education Program of *Time*'s Man of the Year Contest overwhelmingly nominated North as their choice.

As one observer of the hearings suggested, television made the Iran-*contra* hearings a drama and helped transform Oliver North into a celebrity. Reflecting on North's televisual triumph, Moyers said, "I think young people are going to have to be taught visual literacy in the same way I was taught . . . to diagram sentences."[1]

North's guilt or innocence is not the issue here. The point is that, for good and for evil, the tube can powerfully shape viewers' sense of reality.

All television programming, from news to drama, creates a symbolic world for viewers. Every image is a bit of meaning that contributes to our view of the world—especially our attitudes and beliefs about people and events we do not experience firsthand. Unless viewers understand how the tube communicates, they will be naive victims of their own televisual pleasures.

North's attorneys apparently knew how to use the camera for his public defense. John F. Kennedy's image-makers apparently knew how to use it better than Richard Nixon's campaign officials in the televised election debates of 1960.[2] Christians must learn how to be viewers by becoming critical interpreters of televisual images.

The last chapter emphasized the social context for critical viewing. This one looks at how television communicates as a distinct technology as well as a social institution. Once we understand how the tube's images convey meaning, we can critically dissect and evaluate them.

We should recognize that sometimes the most apparently shallow images can affect us the most. Commercials, for example, exude materialistic values, while soap operas are telling enactments of people's anxieties and fantasies. The overall artistic merit of a program is not necessarily commensurate with its cultural significance or raw emotional power.

Why should we become critical viewers? Because part of the responsibility for television rests with audiences. Visually literate viewers not only are less susceptible to the tube's endless stream of images and sounds; they also demand more when they turn on the set, because they understand the medium's potential as a purveyor of culture.

The point of this chapter should be taken to heart by both zealous critics of television and uncritical fans of the tube: The medium needs viewers and producers who are visually literate. Like other visual media, television communicates in its own visual language, or "iconography." In order to redeem the tube, we must be fluent in this language.

The use of words (e.g., dramatic dialog) on television is also very important. Indeed, the intended meanings of particular images are often vague or ambiguous without dialog. But even something as simple as a verbal pause can take on enormous significance when accompanied by images. "No comment" on the tube can have much more meaning than it does in print.[3] "60 Minutes," for example, is full of meaningful images of people—heroes and villains— whose greatness or wickedness is created by camera as well as microphone.

Neil Postman is correct when he says that television expresses most of its content in visual images.[4] But he is wrong when he says that television *must* communicate mostly via image and that television cannot use exposition (rather than narrative) to express propositions or to argue logical points.

Postman assumes that the social institution of television cannot be changed. But viewers and producers can be taught to use the tube to promote discourse and debate as well as rational thought in general. And images, even in narratives, can be used for these purposes.

The problem is that viewers are not taught the language of moving images. Ironically, in a visual society most people are visually illiterate.

The Myth of the Visual Age

It's popular these days to refer to America as a "visual culture." The media frequently address the fact that Americans are very image-conscious. Americans do find much meaning in the purchase and display of visually oriented goods, from automobiles to clothing. Perhaps more than any other culture, and certainly more than most, American life is rooted in images of who people are and who they wish to become.

As sociologist David Riesman argued back in the 1950s, individual Americans are directed less by tradition—an inner sense of the past—than by the expectations of others.[5] In this type of "other-directed" culture, a person's "image" is like a set of clothes that can be worn to impress and to find happiness. Americans are always striving to become something they are not, and such strivings are almost always tied to culturally defined images.

But Americans are also remarkably naive about how images actually function in the culture around them. Like fish in water, Americans take for granted the visual aspects of their lives. Although they feel the pressure of visual style, and often desperately seek to put on the right images, they know little about where the images come from, what they mean and especially how they communicate. Adolescents, for example, feel the enormous burden of winning the approval of their peers, but they may never have thought about who generated the burden or how to escape it without losing face. Similarly, Americans consume enormous amounts of television, video and film, but the vast majority of them have no idea how those media products communicate.

Most of us are visually illiterate participants in a visually oriented society. And the schools do not help us become visually literate.

Iconographic Fears

Strangely enough, the visual impoverishment of the American educational system partly reflects an underlying suspicion of visual communication stemming from Protestant concerns about idols and icons. In the Middle Ages, religious icons "personalized" prayer and worship by directing them (or misdirecting them) to godly personalities, not just to theological ideas or confessional statements. During the Protestant Reformation, reformers tried to correct abuses by ridding churches of icons altogether. Worship in Protestant theology was increasingly equated with the sermon ("the word"), and visually oriented sacraments were relegated to less important roles.

Jacques Ellul's contemporary critique of images reflects a similar veneration of words and disdain for images: "visualizing God's message brought in its wake all sorts of consequences in terms of magic, superstition, idolatry, paganism, and polytheism."[6] Ellul is probably correct that the spread of images began in the church. In some churches the reading of the Word was indeed replaced by "liturgical gestures, colors, changes of clothing, incantations, and litanies."[7]

And image-making spread rapidly—especially in industry. Secular media eventually became the new iconographers, or symbol merchants. But some Christian traditions remained skeptical and even critical of the new visualization, whether in church or industry.[8] Ellul's fears, like those of many Protestants, established a false dichotomy—either images *or* the words.

Meanwhile, the Roman Catholic church, fearing the impact of the print revolution, banned and burned books, "backed up by the Inquisition and a repressive alliance with reactionary political elites." Not until the mid-eighteenth century did the Roman church begin offering print-based papal encyclicals, and one of the major objects of attack was the Protestant Bible societies.[9] While the Protestants feared visual communication, the Catholics worried about printed messages. Each church lost opportunities to enrich its ministries and expand its influence in a period of considerable secularization. Protestants' fears were imported into elementary and secondary schools, colleges and universities. After all, nearly all of the first private colleges in the United States were established by churches and denominations. That religious influence, plus the universities' own scholasticism that emphasized written and printed texts, largely shaped American educators' negative attitudes toward visual languages.

Of course, various kinds of iconography did creep into higher education through the study of the fine arts, especially art history and drama, but even liberal arts colleges were badly impoverished. The Protestant opposition to worship of idols undoubtedly cleansed the church of most icons, but at the same time it left students sorely lacking in the visual literacy needed to understand how the new secular priests and prophets—the electronic media—were shaping the culture's values and beliefs. Perhaps Protestantism's visual impoverishment, evident even in most church architecture, had the effect of creating a greater market for secular images.

In any case, the schools have missed an opportunity to educate about television. As Michael Novak put it, "It seems a shame that educational systems and universities limit themselves to training the literary imagination, and neglect the training of the television and cinematic imagination, in which we are all immersed for so many hours."[10] Whoever controls the stories *and* images of a culture will largely command peoples' values, attitudes and beliefs—and not just in so-called primitive cultures. Modern industrial society has its own idols and icons. Jean Shepherd disturbingly but humorously captures this fact in a futuristic story, "The Lost Civilization of Deli," about anthropologists who uncover the remains of "Fun City" (New York) in underground vaults along Madison Avenue. After examining the filmic images of various television commercials, from McDonald's to Mr. Whipple, the anthropologists rightly conclude that their archeological "find" would help unlock an entire cultural heritage.[11]

Such popular culture is not important for its literary value *per* se, but for its real cultural power. In short, the image-makers and storytellers are more influential than scholars or elite artists. They are the popular poets of the contemporary age.

The Lure of Moving Images
One of the most important aspects of televisual language is the fact that the tube is a *moving*-image medium. The VCR has changed television and film

viewing somewhat. People occasionally stop the videotape to examine an image or scene in more detail, to view an entire program again, to discuss the tape or merely to attend to other matters. But the television show still operates visually as well as aurally at its own pace.

Print media, which are easily adapted to the pace of a reader's own reflections, can more readily foster thought. By contrast, the tube travels merrily along its own path in spite of our desires to slow it down, to think about it or to articulate its impact upon us. This is why the medium can be so frustrating to use in educational settings designed to encourage reflection and critical viewing. It is also why some academic studies of television can isolate people's viewing experience from scholarly interpretation of the same programs. Scholars sometimes "think about" shows entirely differently from how audiences enjoy them.[12]

The shift from live to prerecorded television production during the 1950s set in motion a trend that continues to the present: the use of tight scene editing, multiple cameras, varied camera angles and zoom shots to quicken the visual pace of programs. Programs produced only a decade ago often seem dull when compared with currently popular shows. The difference is not so much in the quality of scripting or performance, but in the visual speed of the scenes—a pace that creates a sense of "action" that could not be duplicated in live studio production.[13]

Except for a few types of programs—most notably soap operas and situation comedies—current shows are visually hyperactive. Most do not hold a particular shot or scene for more than a few seconds at a time. Even variety programs make extensive use of zoom lenses and moving cameras to maximize the visual interest of singers and other performers.

That is a far cry from the two-camera variety shows of the early 1950s, such as the enormously popular "Ed Sullivan Show," which pictured either the entire stage or a close-up of individual performers, with little or no camera movement. The industry learned that the "best way to hold a viewer's attention . . . is by a succession of rapidly changing images none of which is on the screen unaltered for more than a few seconds."[14] As a result, every year American television looks more and more like rock videos. This is true of news and commercials as well as drama.

These developments have inhibited thoughtful viewing. Programs often pull viewers into the story not to communicate a particular message or to establish a proposition, but to get the viewer to feel a particular emotion. In this regard MTV's rock videos and many commercials are actually at the forefront of televisual communication. They are created specifically to hold attention, first, and then to elicit feelings, not to offer or argue ideas. Many televisual messages operate at this impressionistic level—often very effectively. Visual pace and musical accompaniment charge images with seemingly nondiscursive meaning.[15]

Rapidly changing images are not inherently evil or unethical. They are legitimate whenever they actually contribute to the meaning of the program. For example, in an *expressionistic* show such as "The Wonder Years," the grammar of visual pace is nicely employed to create an accurate sense of how the character personally experiences events. When that program's lead character, a young teen, is feeling overwhelmed by events around him, his confusion may be communicated to viewers through the depiction of the outside world as a series of rapidly moving, disconnected sounds and images. Every child has felt this in the confusion of crowds, and especially in a moment of great embarrassment, when the mind speeds ahead of rational thought.

Writer Jean Shepherd used the same expressionistic techniques in his various made-for-television films (such as *The Phantom of the Open Hearth, The Star-Crossed Romance of Josephine Cosnowski* and *Olie Hopnoodle's Haven of Bliss)* and in his enjoyable feature film *A Christmas Story.* The informed viewer should be able to distinguish such legitimate uses of visual pace from the purely manipulative ones designed to hold viewers to the channel without any dramatic purpose.

The revolution of MTV-style programming, which is becoming evident in all types of series and made-for-TV movies, is nothing short of astounding. More and more programming looks and sounds like rock videos and commercials. These impressionistic tales create moods and feelings that usually bypass critical response. Indeed, they are intended to do so in the name of higher audience ratings or greater advertising profits. Even televised sports events are being transformed into high-tech image bombardments through the use of many cameras, faster editing, electronic graphics, replay machines and the like.

The result is programming that communicates emotionally more than cognitively or intellectually. Only when we slow down the pace of some television shows are we able to feel their power and evaluate their control over us. This means turning them off or stopping the tape and engaging our minds.

Some of the most moving rock videos are incredibly sexist, masochistic, violent and even nihilistic. We won't be able to see this fact until we view the images at a pace that allows for reflection. The VCR is an essential tool for critical viewing, because, like a book, it enables us to go back to the same images until we are confident that we can make sense out of them.

Televisual Intimacy

The small screen and poor-quality image define the contours of televisual language. Television is less effective than film for the use of color, lighting and careful visual composition. These are cinematic devices that work well on the large screen, but not on the tube.[16] Fine visual detail is also lost.

For example, in the film *Tender Mercies* Robert Duvall plays Mac Sledge, a down-and-out country-and-western songwriter cast off by his successful former wife, who has become famous and wealthy by performing Mac's material. Mac's personal plight is projected visually in the barren landscape of northern Texas, where he eventually finds love and a new family. In one crucial scene Mac is shown looking contemplatively over the flat, parched land, with the camera behind him and no sign of civilization in his view. The vast sweep of this important scene, which sets Mac for his spiritual conversion, is totally lost on the TV screen.

Television drama moves into the arena of artistic integrity and thoughtful programming when it makes appropriate use of the small screen. The best televisual drama, whether comedy or serious drama, develops thought *and* emotion via dialog and images of the human face. Television is inherently an intimate medium that thrives on character and psychological conflict. As TV producer Lewis Freedman once put it, television can effectively address a "conflict of psychology or character, or a conflict of morality."[17] The tube communicates most powerfully when the human face is the dominant image. In fact, the face is the only image that the tube really captures well. It fits the size of the TV screen and overcomes poor picture resolution. Martin Esslin goes so far as to claim that the "ability of TV to transmit personality is . . . the secret of its immense power."[18] Whereas in the theater the audience sees "everything from a fixed angle—and nothing in detail," on the tube close-ups of faces are key to establishing the characters.[19]

Soap operas and sitcoms are filled with characters' faces. The close-up facial image has become their major technique for registering the agony and ecstasy of the soap and the humor of the situation comedy. News programs similarly spend the vast majority of their time "imaging" peoples' faces, especially the news anchor and the reporters, but also the newsmakers themselves. Commercials register the benefits of products on the faces of satisfied customers. "Reading" these faces is a key to interpreting and evaluating programming, because faces create the persona of character and express the conflict of the story.

On "The Cosby Show," for example, Cliff's facial expressions were just as important as the dialog in interpreting his character. Cosby, a gifted comedian, knew how to convey facially the dramatic essence of the moment. His face *showed* surprise, hypocrisy, frustration, contentment, pride and the like. He never had to say, "I'm feeling superior to my wife," or "Aren't I smart?" because his face displayed these attitudes. The tilt of his head, the height of his eyebrows and the width of his smile (as well as the timing of his dialog and the quality of his voice) all contributed to the meaning of individual scenes and to the overall character known as Cliff Huxtable. This type of facial acting gave "The Cosby Show" a depth of character missing from most sitcoms, which depended more heavily on dialog and slapstick action to generate laughs.

Contrast this with the highly successful "Miami Vice," which was NBC's attempt to create a show based on the idea of "MTV Cops."[20] The series emphasized mood through a combination of musical sound tracks and carefully crafted shots of interior and exterior settings. The emphasis on mood, however, was not as meaningful as the show's approach to character. Except for a few sexual or semi-comedic scenes, "Miami Vice" was filled with the detectives' anxiety-filled faces. Their expressions reflected the essence of the show's meaning: danger, anger, fear, revenge, cynicism. Even though the bad guys were usually caught, the same emotions would resurface in the next show. Clearly, "Miami Vice" was more about despair and anxiety than it was a celebration of police success.

Because cable TV and video markets are increasingly important to the feature film business, more and more movies are being made for the small screen rather than for the big one. The result is movies that emphasize facial shots and are not as concerned about color, the overall quality of set design, the "background" of shots, panoramic shots and so forth. In the theater these

types of films, from *Home Alone* to *Baby Boom,* can overwhelm the viewer
with the full-screen images of faces. But on cable or VCR they look comfortably like other television programs.

Persona

As the earlier account of the Oliver North hearings suggested, television's
image size and resolution accentuate facial images over all others. The basic
symbolic unit on the tube is facial expression. Face represents character and,
along with dialog, establishes persona.

Persona—the public image of a personality—is the most obvious aspect of
televisual communication. The tube did not create human personality, but it
does fabricate artificial ones and distribute them everywhere. Such persona
has good and bad consequences. On the positive side, persona provides for
meaningful drama with believable characters. Stories depend on characters
who act upon motives and cause things to happen.

On the negative side, facially constructed persona deceptively establishes
stereotypes and, worse, personality cults. Consider the case of televangelism.
Some viewers get the impression that all preachers, and perhaps even all
Christians, are hucksters—that religion is merely Bible-thumping, money-
making propaganda. In their minds the emotional pleas of a televangelist are
contrived and deceptive; the persona is an image of fabrication, deception and
perhaps even fraud. Few viewers will forget the tears streaming down the face
of Jimmy Swaggart when he confessed to sinfulness on national television;
however, many people interpreted the image as mere performance.

Other viewers might see the same televangelists as authentic preachers of
the very Word of God. To them, the face is authentic and the persona is the
actual person, not an image. Tears and smiles are not part of a performance,
but honest expressions of the heart of the preacher moved by the Holy Spirit.
So these viewers may pledge their allegiance to particular TV evangelists
instead of to Jesus Christ.

This is only one manifestation of the human craving to follow individuals,
to trust them, to admire them and sometimes even to commit one's life to
them. Television satiates that craving for some people because of its emphasis
on persona.

The combination of television and the human face is psychologically vol-
atile. In the privacy of the viewer's home, the combination is explosive. Tel-

evision thrives on intimacy because, says Esslin, the "satisfaction of the drive to gossip about the experiences of others must be one of the central concerns of all human existence." The tube creates the illusion of face-to-face interaction between individual viewers and the televised personas. Because of the close-up images and the gossipy content, viewers can feel that they actually have a relationship with the people on the screen.[21]

The TV set links viewers with all kinds of personalities and every imaginable type of human emotion expressed on real faces. In the process, American television feeds viewers celebrity role models, voices of authority, images of ideal males and females and many other personality-driven messages.[22] Teenagers look to music and film stars for cues on how to live and what type of person to be like. Males focus on the macho images of muscle-bound movie and sports celebrities, while females gaze longingly at the feminine images depicted in diet cola commercials. Viewers see reflected on the tube the personalities to which they aspire.

To alter Marshall McLuhan's famous dictum, "The medium is the message," we may say that "the medium of television is persona."[23] As Esslin writes, "Television is the most voyeuristic of all communications media, not only because it provides more material in an unending stream of images and in the form most universally accessible to the totality of the population, but also because it is the most intimate of the dramatic media. In the theatre the actors are relatively remote from the audience, and the dramatic occasion is public. . . . Television is seen at a close up range and in a more private context."[24] Along with such intimacy comes naive trust in people whom the viewer has never met.

The Uses of Persona

Few of the millions of viewers of "CBS Evening News" ever met Walter Cronkite. Upon his retirement, various people jockeyed behind the scenes for a chance at his position of wealth and status. Meanwhile, the American press speculated as to who would replace the gray-haired father of television news.

Missing from the speculation, however, was any discussion of how the master's leaving would affect the reports presented on the highest-rated network news show. Americans were more interested in the man than the program, in the star than the news, in the celebrity than the medium.

What was Cronkite's attraction? No one knows for sure, but his fatherly image was certainly part of it. He projected a warm, caring persona laced with sagacity, temperance and patience. Although few of his viewers had met him, they were willing to trust him. Television is flooded with images of news reporters, actors, talk-show hosts, and cartoon characters: Dan Rather, Bill Cosby, Farrah Fawcett, David Letterman, Jay Leno and Mickey Mouse. The tube is so successful at establishing persona that some of its stars have lucrative assets in product endorsements.[25] Actors such as Lee Majors and James Garner have had a steady following since their first dramatic series began decades ago; wherever their images are broadcast, however, the same persona exists. It is difficult for Alan Alda (Hawkeye) and Mary Tyler Moore (Mary Richards) ever to escape their old personas in new roles.[26]

Persona is powerful even on commercials. Mr. Whipple, who made a career out of toilet-paper taboos, and Juan Valdez, the mythical Colombian peasant, are TV creations.

Lisa Bonet, who played the sensible, clean-living teenage daughter on "The Cosby Show" and eventually gained her own spin-off series, "A Different World," boggled the minds of some fans when she appeared in the violent and sensual film *Angel Heart*. Bonet then took her clothes off for *Rolling Stone* magazine. Her response to critics may have really disheartened some of her television fans: "People think you're hot if you're on TV. I don't have a TV, really. I've seen, like, two episodes of my own show."[27] It was not long before her role in "A Different World" was considerably reduced. Her persona had been exposed as an image. Frequently actors are chosen for TV roles not because of their acting skill, but because of their celebrity status and audience-attracting abilities. This produces some of the oddest TV stars. In 1984, Vanessa Williams, who lost the title of Miss America because of her appearance in nude photographs in *Penthouse* magazine, appeared on a soap opera as special guest star. As crazy as it seems, the publicity surrounding her *loss* of the crown had been enough to gain her a place on a daytime television program aimed primarily at female viewers. Moreover, the fact that she had appeared *sans* clothes in a pornographic magazine enhanced her chances for a television role—though the American public is generally disapproving of such activities. Although everyone knew she wouldn't be allowed to appear naked on the soap opera, the publicity surrounding her *Penthouse* photographs was enough to project Williams into public life and onto the tube as a dramatic star. Eventually the same stardom landed her a successful recording career.

"60 Minutes," one of the most successful programs of the 1980s, was an excellent example of how persona becomes narrative character on the tube.[28] In spite of all the awards the show won, and despite its muckraking image, "60 Minutes" was essentially persona-oriented entertainment. Compared with much print journalism, its stories were superficial, sensational and largely unoriginal.

In each twelve-minute segment the program normally offered little more than images of conflict between various people or between an individual and "the system." In some cases, the program was merely a shallow profile of a well-known or at least potentially controversial person: a criminal who may have been wrongly convicted, a flashy but disreputable Third World government leader, a physically impaired person leading a relatively normal life. "60 Minutes" appropriately opened each episode with a ticking clock and flashed images of the various reporters, because the program showed how entertaining an hour-long show can be when it is composed of news personalities narrating short vignettes about other people.

Televisual messages are nearly always linked to persona. They are communicated in and through public personalities. The success of "60 Minutes" was largely the result of the masterful use of persona. Each reporter projected persona, through the magic of television, into tales about other human beings. Such persona was weak journalism, but great television.

Rock videos frequently stress lust by cutting from pulsating bodies to ecstatic faces. The message is "Lust is sexy" and "Sex merits lust"; rock stars become messengers of lust. In this sense Madonna may be the most successful rock persona of the era. She is the new female icon of youth, the star who transformed the restrictions of religion into the freedom of female self-indulgence. Her concerts, recordings, films and videos all proclaim this same message—the story of a generation's alleged sexual liberation. But the symbol of that persona is her face, a contemporary version of the image of Marilyn Monroe, with blonde hair, pursed lips, long lashes and partly closed eyes.

Every facial image on a televised narrative contributes to or detracts from the meaning of the story and its impact on viewers. The most powerful ones are not superfluous or only additions to the dialog. They convey as much meaning as the spoken words. In fact, such images establish the context for interpreting the dialog. When Colonel North appeared before the committee, his image, from the proud tilt of his head to the perfect press of his uniform, crafted a sense of personal veracity, patriotism and righteousness. There was no visual evidence of deception, shame or misgiving. He was proud to be American and sure of his patriotic actions. In other words, the image helped define his persona as a selfless, honest American.

North's television appearance thereby established his character in the drama of the Iran-*contra* scandal, in spite of what reporters and columnists said about him. In fact, more Americans apparently believed the image than they did the words of critical journalists.

When Richard Nixon and John F. Kennedy debated on national television for the presidency of the United States, persona turned the election in favor of the young Catholic Democrat from Massachusetts. In the first truly powerful use of televisual persona in politics, Kennedy's warmth, enthusiasm and confidence overshadowed Nixon's political savvy and experience. Kennedy projected a likable persona, whereas Nixon's images conveyed distrust, uncertainty and perhaps even fear. Kennedy was attractive, while Nixon perspired and looked feeble under the spotlights.

No amount of debate and discussion could alter the essential dynamic of televisual persona. Pundits agreed that the debates on the tube won the election for Kennedy, but not because of the verbal content. More often than not, the medium is persona.

Through the power of the tube to generate persona, viewers develop pseudo-relationships with "intimate strangers." The result is what film critic Richard Schickel calls a "culture of celebrity," in which well-known personas become influential symbols for ideas and use their fame to get across a message or simply to make money. The celebrity controls his or her followers by fostering an "illusion of intimacy." The biggest celebrities, says Schickel, are performers "who appear as themselves, or as what we are gratified to think of as themselves."[29]

In this regard, news anchors and talk-show hosts have an advantage over actors, but the truth is that even the seemingly smallest dramatic role is part of the act of persona between intimate strangers. Without thinking about these types of personas, viewers will never see the images for what they really are—fabricated people.

From Image to Word

The critical viewer of television has the difficult job of translating the tube's images into words. Then the words can be processed by the viewer's mind, evaluated and discussed with other viewers. This is a crucial process that all Christians must engage in if they hope to be discerning users of the tube. The other option is merely to sit in front of the set, hypnotized by the images and helplessly unaware of the waves of emotion elicited by those images and sounds.

In Jerzy Kosinski's novel *Being There,* the character Chance is unable to think about what he watches on the tube. He watches a lot of television, but he can't put the images into a meaningful context in his own mind. The only context he has is the one provided by the tube. So it is with many contemporary viewers, who can neither escape the allure of the images nor make sense of them.

Critical viewing is a thoughtful interplay between viewers and programs. It may not always be deep, significant thought, but it is certainly not just watching flickering images—like the bovine gaze at cars passing on a rural highway.

The motives of program producers can never be known for certain. But viewers should be able to make a relatively accurate assessment of what a writer or producer is trying to communicate in a show. We should be able to tell when there is not much message beyond the emotions generated by rapidly moving images. And we should be able to pull those emotions from our hearts to our minds for thoughtful analysis. In the same way, we need to be capable of reading the faces on the tube, not just for the emotional intimacy they are attempting to communicate, but also for the authenticity of the persona.

A Question of Character

Television's major danger is not that it might promote images over words, but that some institutions can so easily use it to emphasize the wrong images, without artistic integrity and ethical sensitivity. As Martin Esslin has argued cogently, in drama a "complex, multi-layered image predominates over the spoken word."[30] This is the nature of drama, not simply the nature of television. Because dramatic performances show people doing things, they invariably focus on human personality. Esslin suggests that "as a method of communication, therefore, drama is highly effective in conveying *human character,* and much less effective in communicating ideas or abstract thought, simply because in drama every abstract idea has to be incarnated in the utterance of an individual and that individual tends to occupy the foreground and to overpower, at least to some degree, the impact of the ideas he or she is voicing."[31]

Compared even with other dramatic media, television emphasizes intimacy and accentuates character over idea, personality over proposition. And it thrives on the faces of human beings, who are insatiable in their interest in other human beings.

But drama generally, and television specifically, can enhance our understanding of the world and bring great enjoyment. Just like essays, lectures or literature, television can communicate propositions, abstract thought and complex ideas. Television writers can address such ideas indirectly through character, setting, plot and dialog. Or they can do it more directly through documentaries, discussions and the like.

Television can even contribute to public discourse by providing viewers with fictional worlds where people are trying to deal Christianly with the dilemmas of life and the human condition. This type of illumination is entirely possible on the tube. Only our institutions limit what television can accomplish in this regard. As with the stories in other media, televised drama can communicate about our character as human beings—who we are, where we're headed in life, why we were made and what we do to each other. In the 1950s television did this on outstanding programs such as "Playhouse 90," a weekly anthology drama. In one of those dramas Mickey Rooney played Sammy Hogarth, a megalomaniacal TV variety-show celebrity who captures the nation's imagination while destroying his own family and coworkers. That show, now available on videotape, was remarkably prophetic, depicting the impact of celebrity status on dozens of the most popular television stars.

Sammy was made into a hero by the public that watched him. His mythic genius as a comedian was conferred upon him by the viewers. The film *Network,* which in the 1970s won five Academy Awards, echoed the same theme even more compellingly and pessimistically. Dramatized stories such as these are insightful and provoke thought.

Cultivating visual literacy is important for many reasons. Chief among them is the issue of *character*—both the character of the people in the television industry and the type of character communicated through their productions. Indeed, the two are related, for the character behind the creation of programs is ultimately reflected in the characters on TV shows. And nothing attracts viewers more than stories about interesting characters.

Character as Motive

Character is essential in all storytelling, but the difference between the uses of character in moving-image and literary narratives bears special mention. Unless they use a narrator, moving-image media like film and television emphasize what a character *does*. This includes speech as well as all the other things people can do. Literary works, on the other hand, tend to emphasize the thoughts behind the action. Of course there are exceptions in both media, but the tendency is clear, and the distinction is significant for the television viewer.

In the moving-image media motive is rarely plumbed as it is in the novel. This is why the adaptation of many novels, such as those of Tolstoy or Dostoyevsky, is so terribly problematic even for television miniseries, which allow greater time to develop character and plot. Just as short stories and novels are often at their best when they explore motive through the thoughts of characters, television is frequently at its finest when it reveals character through visible action. An image can indeed communicate the equivalent of many words, but words can breathe life into a character's motives with a precision and depth that is hard to match in the moving-image media.

Consequently, the burden on the television viewer to interpret motive and evaluate character can be quite formidable. A medium that seems so easy to watch really calls for perceptive eyes and engaged minds. The benefits of showing versus describing action are largely lost when viewers expect too little from televisual characters.

Christians must view television's personas with an eye as well as an ear for motive. After all, Christians should know how important motive is for revealing the hearts of people.

Conclusion

As a face-imaging medium, television conveys what appears to be the essence of performers and real people. The tube deals in facial images as economies deal in currency. Personas become the medium of exchange in all types of televisual communication. They are the fundamental communicative units that convey the most meaning, the primary visual symbols of the tube.

This should not surprise Christians. The face of a human being, like the face of God, is the most expressive part of us. From our earliest months as infants, we are attracted to others' eyes and mouths as reflections of genuine personage—visual expressions of heart and soul.

Television uses our humanness in the name of all kinds of altruistic and selfish goals. Only visual literacy will help us distinguish between truth and fraudulence in character.

However, critics of television usually focus on the worst aspects of television while ignoring both the medium's accomplishments and its potential. Imagine what even commercial television could be like if visually literate producers with integrity shaped the programming. To put it more strongly, consider what the tube would be like if it were used by gifted Christians to fulfill the Cultural Mandate. Such literate television requires at least three things.[32]

First, viewers must learn to recognize the difference between visual manipulation and dramatic communication. Far too many programs are not really worth viewers' time because they are little more than cascades of rapidly changing images. Unfortunately, too few viewers are willing to turn off their sets or switch channels when the programming is visually manipulative. Certainly rock videos are the biggest offenders, followed by commercials.

Second, visual literacy requires us to exercise our ability to discern the meaning of persona in all forms of programming. We need to read the faces on the tube, interpreting their meaning in the context of each show. More than that, we must remain skeptical of the authenticity of all televised persona, on news programs and commercials as well as drama.

As dramas are played out on faces, so ideas, values and beliefs are expressed symbolically. The medium is persona. No matter how much we like a character, we ought to be shrewd enough to ask what the character really represents before we pledge our political, moral or artistic allegiance. What do Hawkeye, Roseanne, Letterman or Cosell stand for? What motivates them? The essence of televisual tales is normally expressed through such personas.

Third, critical viewing is truly possible only when we translate images into words. If we cannot talk about the meaning of a scene, the message of a show or especially the significance of a character, we are mere watchers. Many programs seem hardly to expect more of us, but we should expect more of what we watch—even for amusement. Watching television without thinking about the show is like driving a car without paying attention to road signs or looking out for other vehicles. Dialog on the tube helps supply our cues for the meaning of programs, but images are crucial elements of televisual messages.

As a technology, television seems like a fairly simple tool for moving light and sound from one place to another. But in human hands, as a social institution, it seems that the tube invariably becomes an iconoclast that competes with religious imagery and challenges religious truth. Perhaps the best we can do is to realize that without critical viewing television will move, under the weight of the Fall, in unholy directions. And the most troubling of these is the adopting of personas as false prophets.

In Paddy Chayefsky's film *Network,* the psychotic former news anchor becomes the "mad prophet of the airways" in his own half-hour sermonlike program. On a set that resembles a small chapel, he rants and raves about the evils in the world while unreflective viewers cheer him on in the studio and at home. Eventually his audience ratings begin declining, and the network executives plot his murder. He has become—like so many real people—a tragic human figure who can no longer find reality. Only as a persona is he worshiped. And only as a persona can he be so loved by a fallen people.

Bill Moyers is correct. We need the schools to teach visual literacy. Either viewers learn to "read" televisual "texts" or they will find that they no longer have much to say about the tube. Worse yet, they might turn out to be mad prophets who say a lot, but nothing that makes much sense.

5
TV CRITICISM/
How to Evaluate the Quality of TV Programs

Finally, beloved, whatever is true, whatever is honorable, whatever is just, whatever is pure, whatever is pleasing, whatever is commendable, if there is any excellence and if there is anything worthy of praise, think about these things. (PHIL 4:8 NRSV)

*L*ATE EVERY FALL *ELECTRONIC MEDIA* MAGAZINE SURVEYS TELEVISION critics across the United States for their rankings of the best shows on the air. In 1991 the highest-rated series was "Northern Exposure," followed by "Murphy Brown," "The Simpsons," "L.A. Law" and "Brooklyn Bridge." The sixth highest-rated program, "Cheers," had appeared on the magazine's list of critics' favorites since the poll was started in 1984.[1]

During the same period, audience ratings painted a different picture. Viewers' favorite program, as measured by audience ratings, was "60 Minutes," followed by "Roseanne," "Murphy Brown," "Cheers" and "Designing Women." The sixth highest-rated show, "Full House," was a darling of viewers but not critics, who usually panned the program for its simplistic plots, predictable characters and shallow humor. "The Simpsons" was tied for thirty-sixth, making it the most popular Fox series but hardly a blockbuster. "Brooklyn Bridge" did even worse, ranking seventeeth and attracting only about 9 per cent of the public.[2]

Not surprisingly, critics and the public do not agree entirely on the shows worth watching. How should Christians evaluate programs? Are there distinctively Christian criteria? Should Christians listen to the professional critics or be swayed by audience ratings?

This chapter suggests that standards are essential for Christians. Without them, we will tend to be swayed by our own laziness or selfishness on the one hand, or by industry hype and social pressure on the other. My own surveys of adolescent and adult viewing by Christians suggest that they implicitly use the same standards as the overall population.

The apparent lack of Christian discernment is not surprising. Selecting which television programs to view can be enormously frustrating. Every year the networks unleash a barrage of several dozen new shows. In 1990, for example, ABC, CBS and NBC alone introduced twenty-two new programs, not one of which was a big hit.[3] Meanwhile, cable channels are now offering their own original series along with reruns and movies. The number of videotapes available for rental continues to escalate. Every year there are more choices—and harder ones.

Christians who feel overwhelmed with the wide array of programming have only a few options available. Three types of nonreligious organizations offer regular advice to viewers hoping to separate the televisual wheat from chaff: program reviewers, social scientists and audience-ratings companies. Each claims to know something substantial about the relative merits of shows. The truth is that these visible critics use invisible criteria that usually don't hold up under scrutiny.

Program Reviewers

Newspapers have long been in the game of reviewing television shows. Some of the most respected dailies, such as *The New York Times* and *The Washington Post,* pay respected and talented writers to preview new shows for the public. There is even a television critics' association, made up principally of lone television reviewers on the staffs of newspapers across the country, along with a handful of syndicated television columnists.

Generally speaking, television reviewers are not well respected by their journalistic colleagues who write about politics, corruption, calamities and other "hard news." At many papers the entertainment writers are among the lowest-paid and least-esteemed journalists. Not surprisingly, few journalists actually want to be television writers. Instead, they often end up there because of a dearth of job opportunities in hard news.

Anyone who reads television reviews ought to know about the realities of the job. Reviewers are usually inundated with promotional materials. They receive stacks of mail: videotapes of soon-to-be aired series, miniseries, made-for-TV movies and the like; sheets of background information on actors, producers, writers and television personalities; real and fabricated interview transcripts with television stars and celebrities (so a reviewer can pretend to have interviewed the person); special "freebies" (from knickknacks to T-shirts) to help a reviewer decide which shows to write about; and descriptions of program plots. It's a major job for television reviewers just to sift through their mail without falling prey to the avalanche of publicity.

In other words, television writers are somewhat at the mercy of the television industry for background information, interviews and ideas about programs. Any reviewer who ignores the publicity material is likely to find it difficult to get future information from networks and program producers. Positive reviews, on the other hand, can result in special interviews with celebrities.

For a number of years I was a TV critic on a half-hour religious program carried nationally on cable. I discovered two interesting things. First, many television production companies would not provide program clips unless the review was positive. They often requested copies of the reviews in advance. Apparently the fact that my reviews were on a religious program was especially troubling to some publicists. Second, some Fox stations refused to run critical reviews of the network's shows—even without clips. We faced this when we began syndicating the program on Fox stations in major markets such as New York. Fox affiliates did not like my negative reviews of "The Simpsons" and "Married . . . with Children," in particular. So much for the freedom to broadcast critical assessments of major TV shows.

Television reviewers are usually caught in a subjective bind: how to distinguish between the objective quality of shows and their own personal taste in programs. Most reviewers never get around this problem. Lacking formal education in the art of television, and forever under deadline, they simply turn their columns into reviews of programs based on personal taste. They liked a show or they did not—for a wide range of reasons. Missing are any formal or consistent criteria for evaluating programs.

Probably the worst offenders among moving-image media critics are the two-person, half-hour movie-review shows on television. As Ron Powers says, the reviewers feel compelled to "slug it out" even if there is nothing to fight about.[4] Dramatic interplay between the reviewers is more important than substance.

Obviously if one finds a television writer whose personal taste matches one's own, this type of reviewing can at least suggest whether or not one is likely to enjoy the program. But reviews based on personal taste are not going to help audiences establish family viewing standards.[5] Christians should be especially dubious of selecting shows purely on the basis of personal likes and dislikes; we all enjoy some things that aren't very good for us, whether it be a prime-time dramatic series or potato chips.[6]

As David Littlejohn has suggested, television criticism is "of a very low order—considerably below the standards set by the better American criticism of film, drama, art, music and literature."[7] Things have probably not improved much since Littlejohn wrote in the late 1970s, except at a few major papers and magazines.

Some people might argue that television hardly merits serious criticism, since there seems to be so little programming of value. But the truth is that there is probably a higher percentage of good-quality television drama every year than there is of novels. Public television alone is enough to keep a good critic busy. Add the better productions of the cable and broadcast networks, and the extent of shows worthy of serious critical assessment is considerable. Americans deserve good criticism, not just thoughtless or puffy reviews.

Scientific Confusion

If television reviewers are generally not very helpful, neither are social scientists, although for greatly different reasons. Worst of all is popular reporting about the findings of psychologists, sociologists and the like. Every few years there is another round of faddish reports about the latest supposedly scientifically proven findings: television makes kids hyperactive, increases viewer aggression, promotes violence, lowers school grades and all the rest. As reported in the popular media, scientific findings become a kind of folk wisdom at best, and cant and superstition at worst.

One of my favorite examples was the widespread reporting that television viewing makes children fat.[8] All the study really found was a correlation between viewing and body weight among kids. It might be that fat kids watch more television, not that the tube somehow makes them fat. Other optional explanations are that heavier kids come from particular sociodemographic groups where people generally watch more television, that heavy viewers burn off less fat because they do not play outdoors as much and that television viewing provides a context for snacking. All of these are possible interpretations for the findings of the study, and it is also plausible that the findings were invalid because of problems in the way the study was conducted.

Let me be very frank about this important area of research: Social scientists do not know for certain the effects of various kinds of programs on human beings, mainly because it's nearly impossible to single out one variable as the cause of social or antisocial activity. Researchers live in a state of well-informed ignorance.

On one hand are thousands of studies that seemingly prove that people imitate television—that televised violence, for instance, begets aggressive behavior. On the other hand are thousands of studies apparently documenting how the tube provides a means of personal catharsis for viewers—that violent programming serves as a release valve for potentially aggressive people. These two theories are clearly contradictory and irreconcilable, yet they exist side by side, each with its supporters.

Yet there is beauty behind the scientific confusion. Created by God as special individuals, as Mr. Rogers might put it, viewers cannot be lumped together as a homogeneous mass of passive videots. At least for the present, social science is not able to deal with the complexity of individual human personality in the context of people watching televised tales. We can listen to the social scientists, but we had best be careful about applying their findings to our own lives and the lives of our loved ones.

My own view is that the majority of evidence comes down on the side of the theory of imitation, especially for children. But social-scientific studies are based on averages and statistical inferences, for which there are always exceptions. Once again, we see that there are limitations to this approach to establishing standards for television viewing. Beware of the latest scientific reports as summarized in the popular media.

In addition, beware of politicians who use hearings and public pronouncements about television's nefarious effects to bolster their campaigns or improve their images with constituencies. Over the years politicians have conducted public hearings on the scientifically verifiable impact of the tube, especially the role of televised violence in promoting aggression among children. The results were not "truth" or "better government regulation of television," but lots of reports often misrepresented by those who held the hearings.[9]

The lesson is clear: Science can easily be used by nonscientists to garner public support and play rhetorical games for power and profit. Don't believe all that you hear, see or read about the "scientifically proven" effects of the tube.

Popularity as Quality

When all else fails, especially in the United States, there are the most popular "critics" of all: the audience ratings. Who has not been fascinated with the weekly Nielsen audience estimates or the annual rankings of the most popular programs? Who has not tuned in a program at least partly because it was so popular? Who has not succumbed to the easy equation that popularity somehow reflects the overall quality of a program?

In democratic nations, audience ratings are an especially appealing way of judging programs. They appeal to egalitarian ideals: The best show is the most popular one. It is as if viewers cast their votes by tuning in their favorite programs.

There is no necessary relationship between the popularity of a program and its overall quality. First, certain television shows are heavily promoted by the networks and local stations. In some cases promotion itself can generate enormous audiences in the short run. David Lynch's wacky series of the late 1980s, "Twin Peaks," was so heavily promoted by the media that it could hardly have failed immediately. Somehow Lynch's publicists managed to secure major stories in weekly newsmagazines, newspapers and TV review programs. Lynch's face and name were everywhere, as were photos of characters from the show. Of course, the bizarre series was not able to sustain viewer interest, but for a time the show was a darling of the critics simply because it was unusual.

Second, viewers select programs only from the options available. A program's schedule is a very important factor in determining its popularity. Sunday night, for instance, is one of the best times. The "Ed Sullivan Show" and "60 Minutes," two of the most popular shows of all time, benefited enormously from their Sunday airings. Friday and Saturday nights are among the worst slots.

Programs scheduled immediately before or after high-rated shows will usually gain millions of carryover viewers. In the fall of 1991, for instance, ABC scheduled a new comedy series, "Home Improvement," on Tuesday evenings between the popular shows "Full House" and "Roseanne." Such scheduling virtually guaranteed a significant audience for the upstart, especially considering that CBS and NBC aired "Rescue 911" and "I'll Fly Away" at the same time. The latter was also new and very interesting, but would have a hard time attracting a following against ABC's line-up.

Third, ratings actually say nothing about how positive an audience is about a show. Often the programs with the lowest ratings have the most enthusiastic viewers—just fewer of them. Over the years programs such as "Star Trek" and "Hill Street Blues" barely managed to survive because of their small but loyal audiences. Eventually "Hill Street Blues" became a major series and helped turn NBC around financially. Yet the program had been remarkably unpopular to begin with.[10]

Fourth, and most important, viewers are usually not looking for the best show. They do not take the time or effort to establish viewing standards. Most viewers can't even express clearly why they like a particular series; they just do. Their own personal taste is hardly a vote for quality.

The new "people meters," which require the viewer to respond on a remote control at least every fifteen minutes in order to be counted in the ratings, are not very helpful. These devices have shown that many viewers are not paying attention to the set while it is on, but they have not measured program quality any more than the old-fashioned ratings systems did. The audience is still treated like a commodity for the sake of the advertiser—like a mere mass of unsophisticated buyers.[11]

Audience ratings necessarily measure little more than viewers' taste at a given point. They add nothing to our understanding of how deeply viewers are satisfied with a program.[12]

However measured, public opinion is only a crude guide to truth or cultural standards. Imagine if the most popular car were deemed the best! Audience ratings are a snapshot in time of what people are watching, not an informed picture of what viewers should be viewing. Sometimes they do reflect quality, as in the case of "The Cosby Show" during its early years, when it was clearly one of the more entertaining, insightful, good-spirited and family-oriented situation comedies on the tube. But the ratings are not a reliable guide, because they are little more than statistical summaries of the public's habits. Many habits need to be broken.

Looking for Real Critics

Fortunately, not all television reviewers are enamored with program publicity, personal taste, social-scientific reports and audience ratings. There are writers such as John O'Connor of *The New York Times* and Tom Shales of *The Washington Post,* who go beyond superficial reviews to more provocative and insightful criticism. They are joined by periodical and book writers such as Michael J. Arlen, Ron Powers, William Henry III, Harlan Ellison and Les Brown.[13]

Viewers can learn much from such talented critics of the tube. Christians should find one or two real critics whom they can depend on for helpful assessments of programming as well as for revealing analyses of the television business. As N. D. Batra has suggested, the real TV critic is a kind of "social prophet" with a "refined sensibility, a sense of history, and a trained intuition."[14]

Until the church produces discerning critics who are more openly Christian, these types of writers are the best alternatives. They share three essential tasks: interpretation, explanation and evaluation. In the process they serve their readers by enabling them to be more informed and critical viewers.

Interpretation. TV criticism should first of all interpret the meaning, popularity and impact of a program or series. In short, the best television critics, unlike mere reviewers, attempt to put the tube in some type of social context instead of merely offering a personal appraisal.

This is an important service to viewers. In fact, it is naive for Christian viewers to tune in programs day after day without any sense of the importance of the shows for culture and society. It is like contributing money to a soup kitchen without any information about whether the kitchen serves healthful fare or a steady diet of soda pop and cheesecake.

Critical interpretation looks primarily at two things: the intent of the program's writer or producer and the apparent social significance of a show. Intent is important; on both public and commercial television writers and producers do make their shows for a reason. It might be to make money or to make a point, to amuse or to illuminate. But viewers might interpret the show differently, so the social significance is not necessarily the same as the

intent of the maker. Lacking shared public understandings of how to interpret
television content, it is not surprising that individuals come to different conclusions.[15] Nevertheless, program producers usually intend to communicate
something.

In the 1970s, for example, producer Norman Lear attempted to depict the
fractures, fissures and conflicts of American society in the show "All in the
Family." From this angle the program's significance was a mere reflection of
the cynicism and generational conflict evident during the Vietnam War. For
many viewers, however, the individual characters, especially Archie or Mike,
became heroes who embodied these viewers' own views. My neighbor at the
time was one of them. He steadfastly supported the prejudices and bigotry of
Archie, who had nasty things to say about African-Americans, Jews, Italians,
women and just about everyone who was different from himself.

Even the most seemingly mindless, trivial programs can have a very important place in public life. "The Cosby Show" reflected the resurgence of American family values during the Ronald Reagan presidency. The show "thirtysomething" captured the ambivalence toward both tradition and personal
freedom among the aging yuppies of the same era. Programs such as "L.A.
Law" and "St. Elsewhere" alternately praised and condemned the new pro-
fessionalism among the same social class, often revealing the conflicts between
occupational success and personal satisfaction.

The most popular evening soap operas, from "Dallas" to "Dynasty," celebrated the American dream of material prosperity and exported that dream
all over the world. Residents of East Germany were watching "Dallas" before
the Berlin Wall was torn down. Is it surprising that after gaining personal
freedom many of these residents wanted first to visit the stores on the other
side of the wall? Those programs' constant depiction of ostentatious lifestyles
was sometimes critical, yet still it lured many viewers who wished to peek
across the economic fence.

These types of interpretations of television programs are obviously subjective. It's impossible to know for certain exactly what makes a program popular, what is the show's theme, what effects a series might have on society or
individuals or even what was the creator's intent in writing or producing a
show (beyond, of course, to make some money). Nevertheless, interpretation
is essential for helping viewers to think about these matters of social importance, beyond personal taste. Critical viewing is partly a matter of interpre-

tation, which is a steppingstone to discernment.

Explanation. Television criticism also requires us to explain factual aspects of a program or series. Every viewer should read regularly at least one critic who displays some knowledge of the history of television, how programs are made, how the industry works and how television communicates differently from film and stage.[16]

This information is difficult to pick up just from reading periodicals and watching the tube. To most viewers, the television industry is a mystery; they don't even know the names of any producers, who are the key people in the industry—far more important than actors and actresses.

The television business is loaded with telling decisions that viewers ought to know about. "Falcon Crest," one of the evening soaps, was conceived by Earl Hamner. Believe it or not, Hamner also created "The Waltons." "Falcon Crest" was supposed to be another family-oriented drama, but the production company had its own ideas about what Americans would watch. Hamner eventually left to pursue other projects.[17]

None of the commercial networks accepted "Sesame Street," which eventually was sold to PBS. Also, "Sesame Street" was overhauled by its producers in the 1970s when a major study found that most kids learned little or nothing from the program. The show was initially conceived as a substitute for pre-school for children from disadvantaged homes, but it eventually became a darling of middle- and upper-middle class parents, who contributed generously to public television stations to keep it on the air.[18]

Such facts, when explained to viewers, should alter the way all of us understand the tube. These explanations can help us become informed viewers, not just passive watchers. Perhaps the most desperately needed explanation today is of the growing divergence of televisual production styles. Historically, television drama was essentially all realism. Characters and situations were depicted as they might look in real life, and stories were plausible, if not always probable.[19] But in the 1980s and 1990s, the tube gave birth to many examples of expressionism and impressionism. "The Wonder Years," as well as segments of "Moonlighting," "thirtysomething" and even "Family Ties," experimented with expressionistic drama that distorted the picture of external reality in order to capture various characters' internal experience. At the same time, televisual impressionism, which used images chiefly to create a mood or impression, appeared on shows such as "Miami Vice" and "Twin Peaks," hundreds of commercials and, of course, many of MTV's rock videos and promotional spots.

We need critics to help us figure out the source and importance of these new televisual styles. One explanation, which I largely discount, is that the television industry was increasingly making shows for the postmodernist culture of the late twentieth century. According to this scenario, the tube finally became a popular purveyor of nonrational, nonnarrative styles of public drama—a folk version of avant-garde art. Although there is probably some truth to this explanation, it flies in the face of counterevidence: that some of the simplest, most straightforward programs were successfully spinning popular tales at the same time. If there is a significant postmodernist impulse in the television industry, it never made inroads in such spectacularly popular shows as "The Cosby Show," "Cheers" and "Roseanne."

A far different explanation is that, for better and for worse, the tube was coming of artistic age. In other words, instead of offering only realistic drama for the traditional viewing public, the television industry was increasingly offering stylistically specialized fare for various audiences with their own tastes and preferences.

"Moonlighting," which attracted a relatively small but very loyal audience, took enormous artistic license compared to mainstream television. Characters spoke to the camera (and thus to viewers), traditional dramatic works such as *The Taming of the Shrew* were parodied, various genres, including comedy and tragedy, were combined and, most unbelievably of all, the program often mocked itself. All these elements, plus impressionistic and expressionistic influences, gave "Moonlighting" artistic depth. Not all viewers liked it, but the program was certainly not a predictable situation comedy, soap opera or detective show.[20] All in all, the artistic impulse behind such a program may be a good thing—at least from an aesthetic standpoint.

The emergence of stylistic diversity on the tube merits viewers' serious interest, although we still need critics to help interpret and explain it all. It used to be that the Big Three networks looked only for shows that would attract large, family-oriented audiences.[21] Even into the 1970s, the tube was still a mass medium dominated overwhelmingly by networks whose aim was for everyone to view their shows.

With the growth of cable TV, VCRs and the Fox network, things changed radically. There will always be a place for a family-oriented hit series, like the

"Ed Sullivan Show" and "The Mary Tyler Moore Show" in their eras, but the irreversible trend is for specialized fare. In the early 1990s the Fox network, for instance, was glad to be the lowest-rated network, because its programming was successfully attracting the coveted teenage audience. By 1991 Fox was the top network for teenage audiences; "Beverly Hills 90210" was viewed by one out of every five teenagers in the United States.[22]

Evaluation. Finally, we need to be capable of evaluating programs and series. Making judgments about the morality of programs is the topic of my next chapter. The rest of this chapter offers a number of evaluative standards that individual viewers and producers can use to judge the relative merits of televisual fare. The same standards can help viewers evaluate and select a television critic for ongoing help in judging the programs they view.

Innovation

As Christians use television to help fulfill the Cultural Mandate, they should not overlook their creative potential. Unlike God, creatures cannot make something out of nothing or simply speak things into existence. But we can produce new cultural artifacts rather than merely copying old ones. Innovative programming reflects this potential for creativity. Within the limits of morality and good taste, Christian viewers should seek out innovative shows that expand the medium's cultural universe.

Unfortunately, commercial television tends to imitate rather than innovate. Whenever a program becomes popular, the networks schedule what one writer calls "copies," "spinoffs" and "recombinants."[23] Almost invariably, these copies are inferior to the original programs. After the rise of "The Cosby Show," for instance, networks scrambled for similar middle-class situation comedies portraying African-American families. One such program, starring comedian Flip Wilson, failed miserably. Another one, "Family Matters," would not have survived except for its wacky break-out character, Steve Erkel, whose whiny voice, obnoxious dress and social ineptness especially amused young viewers. Interestingly enough, Erkel's character was an unplanned departure from the Cosby formula.

There has been a surprising amount of innovation on prime-time television over the years. The so-called ensemble programs, such as "Hill Street Blues," "St. Elsewhere," "L.A. Law" and "thirtysomething," introduced more serious and realistic drama to evening schedules. Situation comedies like "All in the

Family" and "Cheers" pushed the genre's limits for social commentary and character development. "M*A*S*H," for all of its cynicism, charted new dramatic territory on the tube by exploring, week after week, the apparent insanity of war as well as the strange blend of compassion and inhumanity that exists in all relationships. These programs can be criticized on other grounds, but they reflected the creativity of their makers and advanced the cultural possibilities of the tube.[24]

Some of television's harshest critics want to make innovativeness the alpha and omega of evaluation. Steeped in the classicist tradition of the arts, they dismiss existing tube fare as formulaic kitsch and call for fine art on prime time. I believe we should avoid all such elitist rhetoric.

Innovativeness alone is not an adequate basis for judging what one sees or hears. Much of what is produced today by the so-called fine arts community is created for its shock value alone, or for its break with established artistic conventions. The truth of the matter is that innovativeness is one important but insufficient criterion for artists, patrons and viewers.

When innovation is idolized, the need for responsibility on the part of producers and viewers can easily be neglected. Some people have suggested that MTV's rock videos should be lauded for their innovative break with traditional narrative television. On one level that may be so, but other criteria are also important.

Quality of Performance

If God is pleased by our responsible use of innovativeness, so too might he smile at well-done performances. Here the tube too often falls very flat; much televisual acting is simply poor. Actors and actresses are often selected for their audience-generating persona instead of their ability to breathe dramatic life into scripts. Even the best televisual material dies on the dramatic vine when it is not enlivened by good acting. Many of the best stage and film actors and actresses have refused to do regular commercial television service, because the weekly grind of the job can destroy the quality of their performances. Made-for-television films and miniseries are not so artistically oppressive, but regular series, especially the daily soap opera, are the fast foods of the television menu.

Given television's insatiable appetite for dramatic material, every quality performance is a special treat. This is precisely why the Hallmark Hall of Fame dramas have been so consistently well performed over the years. The Hallmark company invests its time and money in special performances that, though inexpensive by feature-film standards, are consistently better than the regular series drama. It takes a particularly (and naturally) gifted actor or actress, such as Bill Cosby or Angela Lansbury, to maintain quality in weekly programs.

Not surprisingly, public television's drama is consistently superior to commercial television's performances. It has maintained this quality partly by refusing to play the series-programming game. Instead it has focused on anthology drama, such as the extraordinarily good "American Playhouse" and "Masterpiece Theater." (Oddly enough, many of public television's finest British programming, such as "The Jewel in the Crown," is purchased from a commercial production company called Granada Television, not from the British Broadcasting Corporation.) Because the best television drama is performed principally through the face, not all actors and actresses can do the job well—even if they have been successful in another medium. The entire body is much more crucial in stage acting than in film or television. So is the voice, which usually must be trained for the live audience in the acoustics of a theater. Televisual performances are idiosyncratic possibilities for displaying a particular kind of creative talent. The performances in programs such as "I'll Fly Away," "Life Goes On" and "The Trials of Rosie O'Neill" were vastly superior to those of "Night Court" and "Empty Nest." We should seek out and applaud stations, networks, programs, producers and performers who care about the quality of their work.

Fittingness

As I argued in an earlier chapter, every medium communicates differently. Television, because of the small screen size and the relatively poor picture resolution, thrives on the expression of intimate drama—drama predicated on psychological instead of physical conflict, and characters whose actions can be expressed via facial images as well as dialog.

Whenever the tube is used appropriately to communicate this type of drama, it rings with dramatic power. In the 1950s, anthology dramas such as "The Comedian" and "Marty," both now available on videotape, fit the tube's inherent capacity for intimate expression. The Comedian traced the megalomaniacal television person represented by the character Sammy Hogarth, a Milton Berle or Johnny Carson of his day. Hogarth, played by Mickey Rooney, turned everyone around him into dependent sycophants and destroyed their personal lives. Marty, on the other hand, portrayed the naive innocence of a man who seeks intimacy with a woman in the face of familial conflict. In both cases, as with many other dramas from "Playhouse 90," the writers, directors and performers knew what types of drama fit the new medium.

Ironically, after additional decades of experience with television, today's commercial industry often displays little understanding of the dramatic fit of scripts and medium. Far too many programs are meant to attract audiences solely on the basis of the names of their stars, the place of the show in the schedule, the rapid pace of images or the incessant promotional hype. Certainly some of the ensemble programs are exceptions. So are the more character-oriented situation comedies that blend humor and pathos, such as the best of "Cheers," "M*A*S*H" and "All in the Family." And there are particularly compelling made-for-TV movies and miniseries, from "The Burning Bed" to "Oppenheimer and Testament," which display a remarkable capacity to develop the intimate potential of the medium.

Nevertheless, there is more drama that fails to match the medium. Action series like "MacGyver" may attract viewers who are tired of comedies, but they don't use the medium very effectively. Even though it is somewhat satisfying to see the hero get out of various tight situations, MacGyver's character is only weakly developed on the tube; he might as well be a comic-book hero. Similarly, "L.A. Law" is far better adapted to the small screen than is "The Young Riders," which is often too cinematic. The best TV Westerns, like "Gunsmoke" and "Have Gun, Will Travel," were premised on psychological conflict and communicated well via the facial image.

In addition to intimacy, the tube has significant potential for continuity. Simply put, continuity is television's capacity for developing character and plot over time, from day to day (as in a miniseries) or week to week (as in some series). Obviously, this capacity for continuity is not inherent in the technology of the tube but rather is a result of the way Americans have integrated viewing into their daily routines. Like a long novel, which can be read in snatches until completed, television can be viewed regularly without all the fuss of going to the movies (as with older Western serials). Now the VCR has expanded the medium's capacity for continuity by making it possible to tape long dramas and view them at one's own convenience.

The result is that the tube can explore issues, ideas and characters at some length. Over the years some weekly series have taken advantage of this capacity by portraying changes in characters, settings and plots over time. "The Waltons" did it with some success, becoming a quasi-historical account of one mythic family during the Depression and beyond. Producer Earl Hamner thought of the series as a novel, with each of the episodes "a chapter in a chapter in a continuing growth and change."[25]

"All in the Family" took viewers through more than a decade in the lives of the Bunker and Spivak family, from the courting and marriage of a daughter, to the empty nest of Edith and Archie, to Archie's final setting in his own bar. Similarly, many fans of "M*A*S*H" realized that the significance of the program to them became connected over the years to the psychological changes taking place among Hawkeye Pierce and the rest of the gang. As characters joined and departed the series, they also entered and left the lives of viewers.

Viewers of "Hill Street Blues" were also treated to remarkable continuity over the years of the show's run. In their own ways, "Life Goes On," "L.A. Law" and "thirtysomething" did the same. These shows had some of the tube's finest moments—in part because of regular viewers' immersion in the ongoing story.

Prime-time soaps, by contrast, have offered little real continuity. Programs like "Dallas" and "Knot's Landing" hooked viewers in a changing plot, but here the use of audience anticipation was more of a marketing gimmick than a means to establish depth of character or to explore significant themes.

Viewers can make important decisions about whether or not to invest their time only in typical series, which revert to the same state of affairs after every episode, or to commit their dramatic interest to shows that explore more deeply and compellingly the lives of the characters. I believe that too many viewers, satisfied only with nightly amusement, overlook the value of viewing continuous drama and discussing it with their friends and family. Whether it's historical miniseries, such as "North and South" and "George Washington," or contemporary lifestyle dramas such as "The Wonder Years" and "L.A. Law," serialized television can enable viewers to look at life and characters with a deeper realism than many people recognize.

Universality of Theme
The tube is able, like stage or cinema, to expand the viewer's philosophical

and cosmological horizons. In other words, television can address through drama the questions that have intrigued thoughtful persons of all times and places. TV need not merely entertain for the sake of audience ratings. It can provocatively illuminate the human condition.

From a Christian perspective this can be terribly frightening—especially to parents who do not want their children's faith challenged by seemingly iconoclastic programs. When television addresses universal themes, it usually challenges existing beliefs or expected behaviors. It illuminates troubling aspects of life instead of merely comforting and confirming. In short, this type of programming raises serious questions and challenges the status quo.

Because of the dominance of television's commercial system, which values audience size and advertising revenues over all other considerations, universality of theme is confined largely to public television. Many of the dramas on "American Playhouse," for instance, examine universal themes in the context of American cultural settings. "Testament," one of its most chilling productions, explored whether or not humankind deserves life, given people's tendency throughout history to take life for granted. A nuclear attack on the United States becomes the setting for looking provocatively at the kind of world adults are creating for future generations. That issue, the value of life, is deeply philosophical and, for the believer, theological. More than a few of my students have never been confronted with the question more powerfully than by this television production.

Another "American Playhouse" drama, "The Star-Crossed Romance of Josephine Cosnowski," addressed the apparent incompatibility of the sexes. The made-for-TV film challenged prevailing stereotypes by depicting women as practical and logical and men as emotional and incurably romantic—all in the context of a seemingly innocuous comedy about a boy's first love. In fact, the drama is an antinostalgic tragedy about flaws in human nature as reflected in differences between males and females. Even if the viewer disagrees with writer Jean Shepherd's views of gender, there is much to be gained by considering this film's message.

On commercial television, universal themes are generally examined so subtly or metaphorically that most viewers, interested principally in amusement, never even notice their presence. Two series can easily make the case: "The Equalizer" and "Gabriel's Fire." On the surface, both programs were little more than entertaining and sometimes violent dramas about do-gooders. In

"The Equalizer," a character named McCall traversed the dirty streets of New York City, helping victims of crime and oppression. In "Gabriel's Fire," a former police officer, wrongly sent to prison for allegedly killing a colleague, was unexpectedly freed by a bright attorney who found evidence for his acquittal.

Many episodes of both series were meant to transcend the usual limitations of the detective genre. In fact, the shows addressed such themes as the nature of evil, the role of forgiveness and reconciliation, the generational transmission of immoral conduct, the place of revenge in social justice and the motivation for doing good works. For the alert and informed viewer, these themes enriched the viewing experience.

The point is not that no show is worth viewing unless it examines universal themes. I don't endorse an elitist view of culture that wants to limit the value of a creative work to its capacity to elicit deep, serious reflections on the meaning of existence. But I do suggest that one helpful standard for selecting and evaluating some programs is universality of theme.

For balanced viewing, we should seek out at least some shows that will exercise our minds as well as our funny bones. When the themes appear to be completely at odds with a Christian world view, we certainly have the responsibility either to turn off the shows or to critically discern the thematic spirits that animate the programs. Fortunately, thematically universal shows are usually not as anti-Christian as most of the programs that provide little more than amusement. Generally speaking, writers and producers who care about addressing universal themes are likely to do so with balance, authenticity and even integrity.

Traditional Aesthetic Merits

In the Western world there is a long tradition of evaluating human creations in terms of aesthetic merit. But unfortunately, there is no consensus as to what aesthetic merit really is. For some people it is the rather unexplainable pleasure derived from a painting, musical piece or dramatic play; for others it is an artifact's formal properties that can be examined and explained. In recent decades some people have tried to expand aesthetics to include financial and ideological aspects of art as well.[26]

Rather than getting bogged down in these esoteric arguments, I can best serve television viewers and producers by offering a rather simple but helpful approach to the aesthetics of the tube. Other things being equal, a program that offers unity, richness and intensity is superior to one that does not. These three criteria can help us judge the overall formal quality of a drama in the context of God's creation. Unity, richness and intensity are universal merits of superior cultural products, whether they are so-called elite, popular or even folk art. We do not usually watch a show only because it embodies these merits, but we are more likely to enjoy and appreciate one when it does.

The unity of a televised drama is simply its sense of oneness and wholeness. When scenes fit together well to tell the story, the drama has unity. It also manifests unity when the characters contribute to the same dramatic purpose or theme. In short, unity requires a show's makers to be on the same course rather than working at cross-purposes.

Unfortunately, many shows lack unity. They seem like a series of scenes and characters without an overall conductor to orchestrate them into one dramatic performance. Scenes and characters come and go without any apparent purpose, except to fill the time (which often is the case because of the half-hour and hour-long program slots that have to be filled). Too often drama on the tube is little more than images and loosely constructed plots. Made-for-television films are probably the worst offenders. They are typically poorly told stories without a dramatic core to hold them together; sex and violence are sometimes the substitute glue.

In spite of the demands in commercial television to produce many episodes of a show in a short period of time, a number of series displayed remarkable unity of theme. The first-person narration on "The Wonder Years," for example, was an excellent device for establishing the young boy's coming of age as the central theme of the series. Moreover, each episode had its own unifying theme about the character's growing insights into adult life, as well as his confusion about childhood. Similarly, "Cheers" and "Life Goes On" often successfully wrapped entire episodes around themes such as forgiveness and self-deception. The programs were not just stories designed to hold audiences, but engaging narratives addressing single ideas or issues. As such, they had greater unity than most television.

The richness of a program is its visual and dramatic variety. If characters are similar and predictable, the characterizations lack richness. If the same dialog and facial expressions are used throughout a program, it lacks richness.

If camera angles are almost always the same (as on the soaps), the show lacks visual richness.

As a formulaic medium with a penchant for repeating successful devices, the tube often lacks variety. Daytime soap operas are often the worst offenders. On the other hand, some 1980s and 1990s ensemble programs ("L.A. Law," "Law and Order") were richly textured dramas with interesting and contrasting characters, settings and images. MTV's rock videos are often incredibly rich, even if they usually lack visual or dramatic unity and are morally repugnant.

Finally, a program's intensity is an important aesthetic consideration. Intensity is always tied to the response the show is intended to elicit. If it is designed to be funny, then the show really should be humorous, and not just seem to be so because of a laugh track or promotional spots trying to persuade viewers that it is "the funniest episode yet." If the program is meant to raise questions about the nature of evil, then it should do so provocatively and compellingly. As a replacement for boredom, television does not have to offer intensity. But as a dramatic vehicle in the kingdom of God, the medium can use unity and richness to provide an intense viewing experience. When that intensity is not in opposition to God's will for our lives, when it offers a redemptive viewing experience, aesthetic merit has served the purpose of the Cultural Mandate.

Over the years one of the most aesthetically compelling comedy series was "M*A*S*H." Based on Robert Altman's sardonic film of the same name, the series rather consistently brought together a rich group of characters for a single dramatic purpose. Thanks to remarkably good scripts and performances, the characters were able to tell their tales with many kinds of intensity—compassion, despair, love, forgiveness, joy and the like. Many episodes were engaging dramas that grabbed viewers early in the story and left them quietly drained or openly exuberant afterward. In other words, "M*A*S*H" was aesthetically superior television, so that it was a memorable viewing experience even for those who watched reruns years later.

Intensity, then, follows the intended emotional or intellectual response to the program. "Cheers" frequently provided intensity, whereas "Who's the Boss?" did not. "Perfect Strangers" overshadowed "Baby Talk" in this regard. Clearly, not all shows will have the same impact on different viewers. Nevertheless, each program will tend to have more or less of a propensity to elicit an intense response from viewers.

Redemptiveness
The standards I've already listed go a long way toward helping Christians separate the wheat from the chaff on television. But none of them alone is sufficient, because they need an explicitly Christian context. I use the word *redemptiveness* to refer to the apparent goodness or value of a human creation in the eyes of God. Television programs can be evaluated in terms of redeeming value even when they are not created by Christians. Redemptive art of all kinds is Christian art. Hans Rookmaaker once put it this way, "Christian art is nothing special. It is sound, healthy, good art, one which has a loving and free view of reality, one which is good and true. . . . Art needs no justification."[27]

It is very important that Christian viewers understand the theological aspects of this concept. The redemptiveness of a program is not just an immediate spiritual impact, such as saving people from their sins. Instead, redemptiveness looks to the sovereignty of God in human affairs and searches out God's goodness and grace in all aspects of culture. In other words, redemptiveness assumes that televisual creations can be a force for good in human culture even when they are not meant specifically to spread the good news of Jesus Christ. Gene Veith, Jr., puts it this way: "God's revelation and moral authority are universally operative even in those who do not acknowledge him. This is the point at which even art by non-Christians can be related to God's message and, to a point, to the work of evangelism."[28]

By contributing to the goodness of cultural life, redemptive programs further the kingdom of God. This is partly why Christian colleges teach courses in literature, film and art history which cover much more than so-called Christian art. Redemptiveness challenges commonplaces about what is Christian and non-Christian art.

Redemptiveness is a slippery concept that is open to abuse. When used too loosely, it can justify nearly any kind of human cultural activity; after all, there's bound to be something worthwhile in all but the most despicable or thoroughly degrading works of art. On the other hand, when redemptiveness is defined too narrowly—for example, as evangelism—it can choke off our enjoyment of art and even limit our appreciation of God's works of goodness in human culture. Martin Marty is correct that we "need more religion in our sitcoms."[29] But we need more than clergy or church settings, or even religious actions on the part of characters. Because God intended us to do more than just preach the gospel and work in church buildings, our view of "redemptive" TV must be broader.

From an evangelical perspective, then, we can say two essential things about cultural redemptiveness. First, redemptiveness is not the same as redemption. Redemptive art provides a fertile cultural soil for spreading the gospel, but it is not the gospel itself. "M*A*S*H" may have set the scene for a viewer to be cynical about humankind, but the show did not actually instruct viewers about how to depend on Christ. Similarly, the series "Life Goes On" clearly expressed the value of human life even among Down's syndrome victims, but it never showed how God revealed his love for humankind through Christ's atoning death.

The theological assumption here is that God's grace can extend throughout the creation. God is sovereign and therefore able to work through those people and cultural products that he chooses. This type of grace does not directly save people. Instead, it shapes the creation in tune with God's ultimate purposes.

Second, the very idea of redemptiveness depends on the existence of the gospel of Jesus Christ. In other words, the biblical story of humankind's redemption becomes the ultimate yardstick for measuring the redemptiveness of television productions. Our biblical understanding and discernment become the tools of evaluation. "M*A*S*H" need not preach the gospel to be redemptive, but its tales must be in harmony with the truths of Scripture if we are to accept it as good and worthwhile in the eyes of God. Similarly, "Life Goes On" was redemptive to the extent that its life-affirming themes did not contradict biblical truths.

Below I offer two scholars' approaches to redemptiveness. The first one, concerned specifically with television, seeks to determine when a program is in tune with a Christian world view. The second one, focused on culture generally, points to redemptive marks that can be found in all human creative activity.

Clifford Christians' Redemptiveness. Clifford Christians suggests that television is redemptive when it satisfies any of three important criteria. First, redemptive television portrays individual human beings as responsible moral agents. In other words, any program which depicts people as neither autonomous nor passive beings, but as active, responsible individuals, reflects something of their real condition before the Creator. According to Christians, whenever a drama or documentary reflects this essential human responsibility, though the program may never speak of the existence of God, it is a redemptive program. He argues that evangelicals can help "stimulate the moral imagination" and engender "moral literacy" by this kind of redemptive programming.[30]

Bill Moyers's extensive work on public television is filled with examples of this moral imagination. But any show that displays even the slightest sense of humankind's inherently moral nature is, says Christians, at least somewhat redemptive. Certainly series such as "Life Goes On" and "The Equalizer" reflected moral agency, even if they could be criticized on other grounds.

Human agency is important because it locates humankind in God's creational norm, not merely in some egocentric context. But as William Fore has argued, "The mass media tend to emphasize man's dominion over nature without man's corresponding responsibility for it."[31] Agency and responsibility are not options. They characterize the reality of human existence in a created and fallen world. According to Christians, redemptive programs capture this truth.

Second, programs that depict the universe as "magnificent and worthy of celebration" are similarly redemptive. Clearly, the tube is not the best medium for this: because of its poor image resolution and small screen, it doesn't communicate the magnificence of sunsets, mountains, skies and oceans as well as the cinema does. Nevertheless, says Christians, whenever the universe is captured on the tube in a way that elicits celebration, there is a sign of redemptiveness—for the creation is indeed worth celebrating.

Among the best examples of this, according to Christians, are nature and science programs, including public television's "Nova." While some might charge that such "secular" shows rarely mention God's handiwork in the creation, Christians believe that by engendering a sense of awe and wonder, such programming can move the viewer toward the frame of mind that opens up the possibility of a God. In other words, when the universe is portrayed as magnificent and worthy of celebration, the program conveys God's transcendence even without mentioning God. As the psalmist might put it, the creation itself praises God and invites creatures to join the celebration. The dismal performance of prime-time commercial television in this regard should be obvious to everyone.

Third, television is redemptive when its programming depicts history as purposive. Shows that give no evidence of development in time, where all things tend to revert back to the way they were at the beginning, are weak in this respect. Archie Bunker of "All in the Family," says Christians, was unable to grow, learn and change; his static place in history rendered the program at least partly meaningless. What is the purpose of history if people and society cannot be changed? Are people simply left, like Samuel Beckett's Godot, to wait for God while they repeat the same mistakes, never learning from them?

The tube offers the potential for continuity and development within series, but such potential is warped when characters repeatedly revert to an earlier state—as is the case with so many situation comedies. In Christians' view, there can be little redemptive about such a state of affairs because it eliminates time, progress and purpose and thus does not reflect the real creation as guided and upheld by God.

Nicholas Wolterstorff's Redemptiveness. As contrasted to Christians' approach to redemptiveness, Nicholas Wolterstorff's view has the sweep of generality and the elegance of simplicity. Although it is difficult to apply practically, his perspective adds an important theological dimension to the discussion. Wolterstorff argues that any cultural products that promote shalom bear the redemptive marks of the Creator. Shalom is a state of harmonious, joyful relationship between (1) human beings and God (delight in God's service); (2) one human being and another (delight in community); (3) the human being and self; and (4) humankind and nature (delight in physical surroundings). Whenever a television program contributes to shalom, it is enhancing human life and furthering the kingdom of God.[32]

What types of programs might foster shalom? Certainly we must include dramas such as "The Comedian" and "M*A*S*H," which illuminate our inhumanity to each other. We might also consider comedies like "The Cosby Show" or "The Wonder Years," which offer insights, provide positive role models and offer a moral context for family life. We should certainly include documentaries about the environment, poverty and the aesthetic blightedness of cities. In fact, it is easy to see how any documentary that sensitively addresses a real social or cultural need could be considered a force for shalom.

Most soap operas and game shows, on the other hand, could not be said to promote shalom. Similarly, inane and negatively satirical comedies, like "Night Court" and "Growing Pains," may do little more than foster viewers' prideful sense of superiority over others.

The difficulty of applying both Wolterstorff's and Christians' redemptive standards should not be taken as an indication that evaluating redemptiveness is unimportant. Perhaps more than anything else, followers of Jesus Christ need to think clearly about their responsibility as stewards and developers of the creation. The concept of redemptiveness is a beginning.

Conclusion

It is time for the Christian community—churches, schools, families and the like—to make television viewing a visible part of the life of faith. Rather than depending mindlessly on popular reviewers, scattered scientific reports and the audience ratings—or, worse yet, depending on nothing but personal taste— Christians must transform their use of the tube into a discerning and thoughtful enterprise.

None of us can do this alone. We need each other to establish practical standards for living with the tube. Either we will use the medium in God-glorifying ways, selecting and evaluating our programs carefully, or the medium will continue to use us.

First, it is important to read regularly the columns of at least one knowledgeable and articulate television critic. The standards provided in this chapter should help readers assess the local television reviewers. If the local ones are not adequate, there are two options: (1) subscribe to one of the better national newspapers (this is often a good idea for general arts coverage anyway), or (2) read books by some of the better critics mentioned earlier. Many of these critical anthologies are available in bookstores and libraries.

Second, viewers should establish standards for evaluating programs. In the home, especially, the question of what to watch is a fine opportunity for building spiritual discernment, encouraging cultural sensitivities and nurturing artistic sensibilities. We need to talk about what we will and will not watch—and why.

Literary critic Wayne Booth rightly suggested that "each of us grapples daily with a barrage of 'art' unmatched in quantity and (potentially) unequaled in range." He concludes that we must respond to this onslaught by developing two "great traditional arts: the art of rigorous selection from the offerings of all comers, friends, hacks and con men; and the art of engaging together in the kind of critical talk that alone can protect us from the selections that are arbitrary and dogmatic." Booth even offers the possibility that by "sharing our grounds for selection, we can create moments that turn even the trashiest offering into a genuinely good time." This is wise counsel.[33]

Too many parents provide children with little more than a grab bag of programs on a restricted list, with little or no reason for the list. That kind of negativism, even when it appropriately makes particular shows off-limits, does little to improve the child's capacity to select and view programs. Many adults try to raise their children with invisible cultural standards. When children grow up, they won't always adopt the standards of their parents. But if they are nurtured in an environment that values standards per se, the children are far more likely to establish their own when they become parents. Third, it's essential that evangelicals focus on the redemptive possibilities of the medium, not just on the evil aspects. In other words, we ought to look for what is good and worthwhile on television. We should share these gems with each other, talk about them and encourage their makers to produce more of them. In this task we ought not to feel alone, for the general dissatisfaction with television programming extends far beyond the cultural borders of the Christian community. Christians should be a source of hope that an invisible God can work through fallen people to make visibly redemptive programming.

The media moguls will continue promoting their lists of the best or most popular shows. After all, viewers like to see what others think are the best programs on the tube. Real quality is far more elusive. But when we find it, we should enjoy it and tell others—including those who made it.

6
BEYOND
MORALISM/
How to Evaluate the Morality of TV Drama

Therefore, get rid of all moral filth and the evil that is so prevalent. (JAS 1:21)

I N HIS REVEALING BOOK *THREE BLIND MICE: HOW THE TV NETWORKS LOST Their Way,* Ken Auletta recounts how NBC decided to run the short-lived series "Nightingales." Producer Aaron Spelling proposed the show to NBC's entertainment chief, Brandon Tartikoff. "I can pitch it to you in a sentence," said Spelling: "Student nurses in Dallas in the summer and the air conditioning doesn't work so they sweat a lot."

Tartikoff responded enthusiastically, "It's a 40 share"—meaning he expected that 40 per cent of the prime-time audience would watch the series. "Let's do it." Tartikoff then invested tens of millions of corporate dollars into the show, only to have it flop miserably in the ratings.[1]

As the major storyteller in North America, television has become an important battleground for morality. In 1991, NBC initially decided not to broadcast an episode of "Quantum Leap" about a military cadet who contemplates suicide after being subjected to persecution for his homosexuality. The show's producer said that NBC was afraid to deal with the subject of homosexuality because some advertisers might pull their commercials from the episode. NBC countered that its major concern was that some young viewer might try to copy the cadet's suicide plans.[2] Eventually the episode was broadcast.

Columnist Cal Thomas, referring to the "sex and perversion" disseminated by the media in the name of freedom of expression, blasted the television networks for frequently addressing sexual topics. He cited episodes of "Head of the Class," "Ferris Bueller" and "Doogie Howser."[3] In the latter, Doogie lost his virginity during the fall 1991 television season—a carefully planned kick-off for another year of profitable episodes. Twentieth Television, the show's production company, proudly announced Doogie's feat in a full-page ad in the trade journal *Electronic Media*. The ad compared Doogie's "first" to the show's first-place standing in the audience ratings. The headline read: "Doogie Did It."[4]

In one sense Doogie's virginity was not important. He is only one fictional character in a sea of celebrities. Yet Doogie's fornication was a public enactment of potentially real adolescent behavior. Teen sex on television becomes a national and even international statement.

There is no neutral ground: every public story has consequences as people view it, listen to it, perhaps think about it and ultimately act upon it. The impact is not usually direct and personal, but broadly cultural. Whether used for amusement, instruction, illumination or confirmation, televised tales contribute to cultural life. Televised morality is inherently political, reinforcing some values and rejecting others. Doogie's actions probably helped legitimize "safe sex" among consenting young people. In an age of AIDS that may have seemed appropriate to many viewers.

The growth of cable and videotape has expanded the scope of morally good and bad programming available to viewers. On the good side, there is far more morally positive fare available for people who are willing to look for it. The Disney Channel and the Family Channel, for instance, are loaded with fine family fare. On the bad side, "adult" theaters have been replaced with a thriving soft- and hard-porn videocassette industry. Michigan's "king of pornography," for example, runs a $75-million-a-year chain of nude-dancing clubs, adult bookstores and other sex-related businesses that thrived during the recession of the late 1980s and early 1990s.[5] Cable channels such as HBO and Showtime bring semipornographic and certainly tasteless movies to tens of millions of American homes. Most video stores carry "adult" movies, which are some of their most profitable merchandise.

This chapter ventures into turbulent waters: the morality of television programming. Christians and Hollywood have long fought over filmic and televised portrayals of sex, violence and profanity. In fact, moral outrage frequently brings together disparate churches and denominations for a unified condemnation of offensive movies and programs. Morality in Media and Donald Wildmon's American Family Association have done this.[6] Christians understandably feel like a beleaguered group of viewers trying to defend public morality in the face of repeated assaults by vulgar media moguls.

Unfortunately, viewers—including Christian viewers—have to share the blame. We are part of the problem. Film writer and director Paul Schrader puts it this way: "When you write movies you write for mass audiences, and the two fantasies which permeate all levels of society are sex and violence. You know that if you address a movie to the fantasy of sex and violence that it will appeal to everyone of every social strata in the entire world."[7]

So-called erotic or adult movies now represent between 10 and 20 per cent of all video rentals in the United States.[8] Americans rent and purchase more adult tapes than exercise tapes, music tapes, sports tapes, how-to tapes or classic film tapes. The only tapes that are more popular are children's and new releases. A nationwide survey in early 1990 found that 68 per cent of video rental shops carried sexually explicit tapes.[9] There is no question about it: sex sells.

Network directors know that sex and violence are among the most effective means of luring viewers to the tube. The ratings have proved this point repeatedly. Barry Diller, Fox's former chairman, said in 1989, as the network was beginning to challenge the big ones, that the highest-rated segments of its "The Reporters" series were also the ones that people were most squeamish about because of blood and sex.[10]

In fact, networks and local stations increase sex and violence during the quarterly ratings sweeps, when viewers translate into advertising revenues. Those are the periods when we get the most lurid miniseries and made-for-television films, as well as news shows that voyeuristically highlight the more outrageous activities in our communities. Under the guise of serving the public, reporters tell and show us about the constitutionality of nude dancing as freedom of speech, the horrors of domestic murder and the affluent lives of big-time drug dealers. It is not always clear whether viewers watch in order to learn or to be titillated and entertained. Too often our moralistic outrage over televised immorality covers up our own deep curiosities and fascinations.

Christians, too, are fallen viewers. In 1981, when the Moral Majority was in the media spotlight, ABC commissioned a study of the viewing habits of the organization's members. Obviously, the network was a biased source. Nevertheless, it is not surprising to me that the study found that less than half of the supporters of the organization agreed with attempts to influence program content to conform to their own standards and values, and that Moral Majority supporters apparently watched as much "immoral" television as non-members. For example, 34 per cent of them watched "Dallas," compared with 36 per cent of the general population.[11] My own informal surveys of church members and students at Christian schools would strongly support such claims. Evangelicals are attracted to the same televisual techniques that lure other viewers.

Christians have a legitimate gripe with the television industry, but they will never significantly improve the quality of programming until they clarify their own moral agenda. Many times our condemnations of the industry are misguided or misinformed salvos—a kind of moralistic shooting from the hip. It makes no sense, for instance, to write letters of complaint to networks about programs or films of which we have no firsthand knowledge. Yet this is precisely what some watchdog organizations encourage us to do. Sometimes they even send preprinted postcards for supporters to sign and mail to networks and advertisers. These efforts are a kind of orchestrated propaganda.

We need to be far better attuned to the breadth of our moral concerns. Morality covers far more than the typical issues of sex, violence and profanity. Even the networks know this.[12] In fact, Christians' overemphasis on these three issues has helped to stereotype all believers as puritanical prudes who never enjoy life and who believe the flesh is inherently evil. Unfortunately, television watchdog groups contribute to this negative stereotype.

Sometimes Christians major in minors and minor in moral majors. One instance of explicit sex can send us to the typewriter, while recurrent racism or deeply rooted materialism may not cause us any concern.

For another thing, Christians ought to concern themselves with more than the morality of programming. Previous chapters addressed some of the other standards we should invoke in making judgments about television fare. God expects far more than morally unobjectionable fare on the tube. Indeed, we cannot seriously fulfill the Cultural Mandate and the Great Commission with old-fashioned situation comedies and moralistic television movies. If only the morality of TV is improved, we might end up with a lot of inoffensive but humdrum programming.

It is time to go beyond moralism as we redeem television. Perhaps poor-quality television is itself a moral issue.

The Big Three

Christians are often rightly concerned about the "big three" immoralities—sex, violence and profanity. But why are other moral concerns so rarely mentioned by evangelicals? And how "big" should these three concerns be in our selection of programs for adult and family viewing? Are these always the most important moral aspects of televised drama?

Generally speaking, evangelicals' concerns about sex and especially violence on the tube reflect a long-standing American preoccupation. In many European nations, sex and violence are not nearly such sensitive issues. Sex and violence in novels and short stories have elicited public concern in the United States for over a century, and a few vocal critics continue to press local school boards to censor various books from libraries and classrooms. Not surprisingly, then, Catholics and Protestants alike were sometimes deeply critical of the apparent immorality both in the Hollywood community itself and in the movies it made.[13]

From the beginning of television in the 1940s, the public issues were usually the same. Sex and graphic violence, along with a few expressions of outrage about "blasphemy" in religious dramas, dominated public concern. The U.S. government responded with a number of scientific studies of the impact of violence. In part, these studies were politicians' responses to the concerns of their constituencies.[14] All they really accomplished was to further politicize moral concerns. Social science could not answer moral questions.

American evangelicals' recurring preoccupation with sex and violence, and to some extent profanity, reflects a distinctly American sensitivity to morality. Americans tend to equate morality with personal conduct—especially sexual practices. In other words, individual morality and personal piety are parallel aspects of the American cultural milieu. Conservative Christians share their nation's penchant for taking a very personalistic approach to morality. This is not all bad, since morality—our sense of good and bad, right and wrong—

is partly a matter of individual action.

However, by focusing on the big three immoralities we tend to become blind to societal good and evil. If morality is only a matter of personal conduct, there is no reason to judge the values and practices of society. And, in fact, we rarely hear Christians complaining about televised racism, for instance. Why were evangelicals not at the forefront of public criticism of early television's "Amos 'n' Andy" series, which was eventually taken off the air for its stereotypical depiction of African Americans?[15] Arabs, Chinese and especially native Americans have been savaged by popular television. Jack Shaheen writes that the tube perpetuates four myths about Arabs: "They are all fabulously wealthy; they are barbaric and uncultured; they are sex maniacs with a penchant for white slavery; and they revel in acts of terrorism."[16]

Ironically, one of the Christian cable networks was a leader in recycling old Westerns in the 1980s, yet those programs were the worst examples of racism directed against Native Americans. When the tube confirms society's collective immorality, we are less likely to become outraged or even to complain.

Morality cannot apply to the personal sins that preoccupy evangelicals or any other social group. In the fullest biblical sense, immorality includes such collective sins as racism, materialism, ethnocentrism, sexism and nationalism. Christians should not stand up merely for the morality of their personal lifestyles, but more broadly for matters of truth and justice that affect all of society.

It is a sad fact that evangelicals complain loudly about the televisual stereotyping of believers while ignoring the way the tube glamorizes some of the worst institutionalized sin of contemporary society. Ron Powers has suggested that "Lifestyles of the Rich and Famous," though a seemingly benign show, has "plucked open America's soul and read the price tag."[17] Christians bark about the loose sexual mores among characters in prime-time soap operas while completely overlooking the glamorization of materialism in programs such as "Dallas," "Dynasty" and "Knot's Landing."

In her study of watchdog groups (also called pressure groups, special-interest groups, lobbies and citizens' groups), Kathryn C. Montgomery found that these organizations' narrowly construed concerns frequently made them ineffective. To be really effective, a watchdog group has to frame its concerns in terms of the existing view of public interest, not just in terms of the group's preoccupations. For example, gay activists had more success influencing network portrayals of gays when the society was more open to liberal social mores and less success during the more conservative 1980s.[18]

If evangelicals do not heed this lesson of history, they will waste enormous resources on impotent moral campaigns to clean up television. It's quite likely that network executives see evangelical watchdog organizations as just another pressure group, along with minorities, women, gays, seniors, the disabled and social-issue groups such as the Population Institute and the Solar Lobby.[19]

There is no doubt that television can legitimize immoral behavior simply by broadcasting it. The tube can desensitize viewers to immorality while creating a public sense that morality is only a matter of personal values. This is precisely why Christian viewers should not let their sense of morality be limited to the stereotypical complaints about sex, violence and profanity. The big three surely must concern us, but never to the point that we let ourselves be stereotyped as humorless prudes or one-issue hecklers of the networks.

In short, we need to keep in mind that morality is more than personal behavior, and moral criticism is far broader than moralistic reaction to the big three. All programs we view, including the inoffensive ones, ought to undergo serious personal or family scrutiny for their moral vision of life.

Portrayal vs. Point of View

One of the major Christian-oriented watchdog groups evaluates television programs almost purely on the quantity of sex, violence and profanity. It counts the number of violent acts, for example, and ranks programs according to that statistical scale. In one sense this is a valuable service to viewers who don't have time to watch the programs before deciding which ones are acceptable. In another sense, however, it's a dangerous means of evaluating the moral vision of a program.

This watchdog group, like Christians generally, fails to distinguish between the *portrayal* and *point of view* of televised drama. Consequently, it promotes the naive idea that the simple quantity of sex, violence and profanity is the measure of a program's morality or immorality.

Consider the case of Fox Television's popular "Married . . . with Children." The program certainly had a greater-than-average degree of sexually oriented dialog and situations, and it used more profanity than most situation comedies. On the surface, this would make the program unacceptable to some

Christian viewers. But the greatest problem with "Married . . . with Children" is not its portrayal of sex and profanity, but its point of view *about* sex and profanity. In fact, the program's view of sexuality was not all that different from television's view in general, described by two researchers as "bawdy, humorous and exploitative."[20] Given the program's overt portrayals of such a point of view, it was not surprising that one viewer personally launched a national campaign against the series.[21] Nevertheless, one might wonder why one show elicits public discussion and another with the same point of view escapes scrutiny by media and Christians.

This series was primarily a mockery of both the old-fashioned sitcom and the traditional family. It viewed the middle-class family as a rather sick institution that largely fails to meet the sexual needs of its members. The husband cannot satisfy his frustrated wife, while the daughter's overactive hormones drive her toward juvenile nymphomania. "Married . . . with Children" is a far more morally repugnant program than suggested by the statistics of its portrayal of sex and profanity. The show pretends to be a semiserious satire on sitcoms, but it really views the American family with considerable disdain. Its more tender moments express only a silly sentimentalism that serves to satisfy otherwise unhappy viewers.

A program's obvious portrayal is a legalistic moral yardstick, while its point of view establishes a moral perspective. When a program is loaded with offensive scenes, it is probably not worth viewing. But the real moral "meat" of the show is the overall point of view taken by its makers.

Television dramas do not only depict people doing things; they also usually make implicit judgments about the morality of those actions. Two shows may portray adultery, for example, from different points of view: One celebrates it while the other condemns it.

By overlooking this aspect of television drama, Christians have lost the basis for informed public criticism of programming. Christian criticism of the tube almost always comes across to the wider society as knee-jerk moralism. And children who grow up in such a moralistic environment typically get the wrong message—namely, that their sexuality is evil or that their parents are opposed to sexuality per se.

Surely profanity is a different matter. Unlike sex and violence, it cannot be readily justified by a show's point of view. Still, the Christian community generally fails to consider a program's point of view in making judgments about minimal profanity—even in the case of an otherwise excellent drama. A few years ago some friends semi-jokingly criticized me for recommending that they rent a videotape of a movie with some minor profanity—*Tender Mercies*. They were offended by the portrayals, even though the story's overall point of view was obviously Christian. The profanity gave clues to the main character's feelings about God before his eventual conversion to Jesus Christ. I regret having caused the offense, but I am also saddened by some Christians' inability to understand point of view.

It's impossible to grasp the message of a program without considering its point of view. No program is just a set of depictions of human actions; every one puts the actions in some thematic context. Detective programs, for instance, can glamorize evil in the context of bringing "bad guys" to justice. "Miami Vice" is a good example. More than a few contemporary television series have created virtueless "good guys" who repeatedly use violence and break the law in order to stop violence and uphold the law. Such inconsistencies of moral vision show how important point of view is in interpreting and evaluating programs. The idea that ends justify means is deeply embedded in prime-time drama.

The Limits of Portrayal
As we consider point of view, we must also admit that even when the overall message seems to justify some sex, violence or profanity, there should still be limits. A satisfactory point of view in a show should never be an excuse for unbridled depictions. Three criteria can significantly help us to judge portrayal, even in the context of a drama apparently justified by point of view.

First, it is very important to consider the *necessity* of the moral depiction for the overall message of the program. I admit to liking ensemble programs, from "Hill Street Blues" to "L.A. Law"—prime-time, multi-cast dramas that follow the personal and professional lives of characters in related subplots. In terms of overall artistic quality, these shows are among the best series that have appeared on the tube. But much of the sex, violence and profanity on these types of series is unnecessary. They reflect writers' and producers' tendency to push the limits of what they can depict and still get by the network censors or advertisers.

Toward the end of its network run, "Hill Street Blues" even showed a man baring his buttocks as a statement to someone else. That kind of sensationalism gets publicity for a show, but it is merely gratuitous. The same statement can be made in a much less offensive manner through dialog.

Second, moral depiction should be evaluated on the basis of its *explicitness*. As a general rule, the more explicit the sex, violence and profanity, the less artistic merit in a program. Pornography can be defined largely by the explicitness of its depictions. The greater the explicitness, the more wary we should be of a show.

"M*A*S*H" did not need to show a lot of explicit violence in order to advance the point of view that human violence is a tragedy, even in war. Graphic violence is not a sign of merit, even when the point of view seems to justify violence in the name of justice, self-defense or some other moral value. Nor are explicit depictions of sexual acts easily condoned on the basis of a program's positive theme.

We do not know the long-term impact of viewing explicit scenes, although it likely desensitizes people to the evil of the actions depicted, so that they become rather matter-of-fact activities. There is no good reason to watch these kinds of scenes, which are usually attempts to attract desensitized viewers with ever greater stimuli. In its day "Gunsmoke" was considered quite violent, and many parents would not let their children view the show. Thirty years later, "Gunsmoke" is extremely mild compared to many movies that are delivered to homes on cable in the middle of the day.

Third, we should consider whether all portrayals of sex, violence and profanity are *appropriate* for a particular audience. This is especially important for broadcast television, which is available to almost everyone who owns a set. Usually we think more carefully about subscribing to cable—and especially extra movie channels—than we do about making regular over-the-air television available in our homes. In either case, however, it's clear that not all shows are acceptable for all audiences. Young children can be greatly confused and disturbed by television programs that address adult themes. This is partly why the commercial networks have complied with the Federal Communications Commission's attempt to maintain some type of family programming during the first hour of prime time.

The expansion of television to cable and the VCR makes it even more imperative that we judge the appropriateness of programming for various audiences. Today it is far less likely that a given program or film will automatically be appropriate for all viewers.

Rating-code designations can help somewhat for movies, but they are only a crude measure of the explicitness of the sex and violence and the amount of profanity. And the film code is virtually useless for helping us judge a movie's moral vision; a film rated for a general audience can have a despicable view of human life. To judge appropriateness, we have to become better informed about programming and its influence on people.

Audience Susceptibility

In its attempt to shape the moral fabric of American entertainment, and to defend itself from the negative impact of evil programming, the Christian community too often has overlooked the fact that individual viewers vary in their susceptibility to moving-image stories. The same program can affect individual viewers differently.

This is a fascinating aspect of being created in the image of God. Each of us is a distinct personality, both because of the genes we inherited and as a result of the way we were raised. Frankly, I don't fully understand why the same program or movie will contribute to one person's moral decadence while being apparently harmless for another person. This difference is a mystery for everyone, including the psychologists who sometimes try to explain it.

I'm not calling for spineless tolerance or a morally repugnant relativism. Instead, I ask all Christians to consider the implications of our createdness for the overall Christian community as well as for our individuality. The tension between personal Christian freedom and the obligations of Christian community bears directly upon our television viewing. With the advent of cable and videotapes, it is even more crucial that we address the obvious differences in personal and societal aspects of televisually portrayed morality.

As a parent and college professor, I have seen the influence of peer pressure. At all ages people tend to feel that they should watch particular programs because others are viewing them. During the adolescent years, this is an extremely powerful influence on viewing. Because of their need for self-esteem and social acceptance, most teenagers find it hard to reject the collective values of their peers. Children and adults, too, sense that their viewing habits will reflect their image to others. Some individuals feel pressured to conform their viewing to the more tolerant standards of their primary social groups.

On the other hand, without sharing the trust and discernment of other people, each of us can too easily decide that television viewing is just a matter of personal taste. I believe the adult Christian community now errs primarily in this regard. We rarely talk seriously with each other about what we're watching and what impact the programs may be having on our moral, spiritual and artistic lives—for good or bad.

I am astounded at the number of spouses who rarely talk with each other about the shows they watch. It is as if television viewing exists outside of their conscious relationship, as well as outside of their church community. Eventually this type of individualistic attitude is passed on to their children, who learn by example that the tube is little more than an entertainment box for satisfying their own needs for amusement, pleasure and the like.

Our real differences in moral susceptibility must be respected in the context of Christian community. In other words, both the individual and the group have rights and responsibilities. We need to allow individual viewers to formulate personal viewing standards, but never to the extent that they greatly violate community standards or lead to personal harm. The number of people who are secretly addicted to video pornography is probably much higher than any of us would imagine. And many people who avoid pornography are still damaging themselves by addictions to soap operas or harmful cable-TV movies.

At the same time, we should not let the prudes and moralists decide on morality for the entire community. Similarly, it's wrong for individual parents to subject their children to narrow-minded, uninformed standards. Parents need to make sure that they know what they're talking about, and that they don't simply pass along their own insecurities and moral anxieties to their offspring.

Televised sex and violence can affect even children from the same family very differently. For one child, sexual portrayals may be a source of erotic stimulation, while for another one they may be little more than a boring part of the story. Similarly, one child's television-induced aggression may be another's catharsis. Families cannot face these differences intelligently until they identify and accept them as part of humankind's created individuality as well as its fallenness.

Art vs. Morality
Evangelicals often reject the confusing distinction between the aesthetic and moral aspects of televisual entertainment. Instead of accepting the separateness of the two, they try to collapse the aesthetic dimension of the arts into

the moral dimension. A program is either good or bad, clean or dirty, moral or immoral. In this perspective, a show can never be aesthetically good and morally bad, or vice versa. Morality defines television.

As Gene Veith, Jr., has argued, "Although the relation between content and form is very complex, the two are in principle quite distinct." He adds that we may aesthetically appreciate a work of art without agreeing with its content.[22] According to Kenneth Myers, Christians must recognize that "aesthetic 'goodness' is not the same as moral 'goodness'—that the word 'good' in the phrase *good art* is not a moral evaluation."[23]

Unfortunately, Christian television is usually characterized by morally inoffensive but aesthetically inferior programs—poor-quality imitations of secular fare. For example, Christian variety shows avoid performers with undulating hips and low-cut dresses—but talent is sadly lacking. Singers on Christian TV claim to be performing for the Lord, but only the lyrics give such an indication. The children's cartoons are usually strikingly below average in technical quality, as are their story lines and characterizations. Christian game shows, usually involving Bible "trivia," are embarrassingly poor productions.

Christian television just does not compare artistically with most prime-time fare. Surely a lack of finances is part of the problem. More important, though, is the mistaken notion that morally inoffensive programming is all the Lord expects of his people in a television age.

Once we recognize that a show can be good or bad both morally and aesthetically, we are well on the way to redeeming our viewing habits.[24] God expected the temple to be built as an aesthetically pleasing structure, not merely as a place for morally upright activities.[25] The Cultural Mandate extends over all aspects of human culture, including the aesthetic dimension of life. Humankind is not called only to make culture, but to make pleasing and attractive artifacts of the highest quality possible. There is no place for cheap, tawdry productions, even if they seem morally righteous.

This is not an elitist argument against popular art and in favor of fine art. The idea that television drama should only be Shakespearean productions or Italian opera is misguided. Stories have many different functions, as I argued earlier. Even amusement, the most maligned use of narrative art, has its legitimate place in human culture.

The central question is this: Are we making and viewing the best programs for amusement, pleasure, illumination, confirmation and instruction? In other words, are the caretakers of the creation being careful stewards of their creative gifts and earthly resources? Inoffensive programs are not enough. God expects well-crafted shows, whether they are comedies, dramas or any other type.

By "aesthetic quality" I mean the general artistic merits of a production, not just the formal aesthetic criteria discussed in the previous chapter. This includes how well a drama is written, directed and edited, as well as how masterfully it is performed. We forget that how well something is crafted is often a moral issue. Someone has to decide what is the *right* thing to do—the proper way to direct, write and so forth. These decisions are based on what the makers *value* in their work. Are they seeking mainly to please the boss, to do as they are told to ride up the corporate ladder, to please God or to impress colleagues? Choices such as these are invariably ethical, not just professional.

Imitation of secular products is not the answer to God's call for aesthetic excellence. Too many Christian viewers look for religious programs that merely copy the secular shows. We should look for the best programming possible, not just the most popular or familiar. Clearly this will not always be an easy task, considering the secularity of public standards.

Christians cannot dictate public taste in a pluralistic society, but they surely should work with others to shape government policy and commercial broadcast practices. Over-the-air television is explicitly regulated in the United States by the federal government to meet the public "interest, convenience and necessity." Is the public not better served by quality fare?

Unfortunately, Christians are reluctant to give credit where it is due for improvements in televisual quality. Numerous Christian groups blasted producer-writer Norman Lear for the insensitive ways he addressed some sensitive issues in the 1970s on "All in the Family," "The Jeffersons," "Maude" and other comedies. But did we ever thank him for reviving the ailing sitcom as a vehicle for social satire and cultural criticism? His characters and stories were far less moralistic and much more realistic than those of earlier comedies. And what of Stephen Bochco's work on "Hill Street Blues," "L.A. Law" and even "Doogie Howser"? Regardless of some of the specific moral content of these programs, they were far more creative than most TV fare. Is it not possible for us to appreciate the artistic talents behind such programming without condoning all the portrayals and points of view represented in each episode?

Among the strange ironies in the legislating of televisual morality is many Christians' steadfast support for commercial broadcasting. The commercial broadcast system is most likely to air morally offensive shows simply because such programs often are among the most profitable. Free-enterprise broadcasting has consistently shown a low regard for morality and good taste, primarily because viewers want instant entertainment and don't like to *think* about television. Commercial television encourages people in the industry to pander to the mass market for the love of money.

The Christian community fools itself when it seeks purely inoffensive programming without regard for aesthetic quality. Among all types of shows, from serious public television dramas to commercial network comedies, Christians should choose programs that are aesthetically superior.

Legislating Morality

Christians who care about the moral dimensions of television programming face the difficult problem of how best to influence the content. After all, Christians should care about what is aired or transmitted on cable, even if they and their immediate family do not watch it. The tube is one of the most public media in the nation and, increasingly, in the world. Just as believers have a responsibility to ensure the best possible schooling for their communities, they should look at ways of promoting artistic quality.

Perhaps the easiest way of doing this is to become supporters of public television. Because of reduced government funding of public television in the 1980s, local public stations have had to turn to corporations and individual citizens for donations. Christians should be among the biggest supporters of noncommercial programming. Beyond "Sesame Street" and "Mr. Rogers' Neighborhood," public television consistently produces some of the finest and least morally objectionable fare on the tube.

Commercial television is a different matter. Tens and perhaps hundreds of thousands of Christians have participated in economic boycotts against sponsors of morally offensive programming. Evidence on the effectiveness of such boycotts is mixed,[26] but it is an acceptable way of bringing complaints about programming to the people who are most likely to respond. This is not censorship, as some critics suggest. The networks still have the right to air whatever programs they wish, and advertisers have the freedom to sponsor any programs.

But Christians have to be careful that they are not in the business of trying to impose their own *taste* in moral programming. Morality can never be totally regulated, and taste is a matter of personal preference, not moral absolutes. In a pluralistic society that embraces freedom of speech, personal morality cannot be legislated.

Christians should focus on the outrageous abuses of the freedom of broadcasting—unacceptable abuses overlooked by the Federal Communications Commission and of general *public* concern. For example, the U.S. Congress had the sense some years back to ban cigarette advertising on television. As we now know for certain, smoking is a major social health problem. Similarly, Christians should object publicly to programming that has little or no social value and that could be significantly harmful to people's emotional or physical health.

Matters primarily of personal taste, on the other hand, should not be subject to law. As long as evangelicals complain only about religious stereotyping or sexual innuendos, for instance, they will have little public credibility. Such complaints usually sound like the grumblings of special-interest groups, from gays to feminists, who hope to legitimize their values and practices by intimidating people in the television industry.

The fine line is somewhere between public decency and personal taste. When a program offends decency, Christians ought to join the chorus of public protest. The network broadcast of the film *Helter Skelter,* in my judgment, violated even a minimal standard of public decency. What possible merit was there in airing on national television a dramatization of the bloody, cultic slayings of innocent people by Charles Manson and his followers? An informative documentary about the dangers of such cults would probably have been acceptable, but there was little legitimate social purpose behind the filmic re-enactment of the heinous crimes. This is not simply a matter of personal taste. In such cases the public has a responsibility to complain loudly and persistently to the offending network and to the FCC, as well as to the sponsors of such trash. In one of the more despicable cable television examples, television covered the live trial of a group of men charged with gang-raping a Massachusetts woman on a pool table in a tavern.[27] This kind of moral garbage should never be delivered to people's homes.

The Need for Public-Issue Programs
Christians' concerns about the morality of television should never obfuscate

the crucial need for drama that addresses serious public issues and needs. Unfortunately, some evangelicals associate such programs with immorality. Any show that deals candidly with delicate issues such as homosexuality and incest is quickly criticized—sometimes even before it airs.

That happened with the network movie *The Day After,* which addressed the impact of a nuclear attack on a Midwestern city. Fearing that such a program would play into the hands of liberal foes of nuclear weapons, numerous public figures blasted the show and the network, and eventually most sponsors pulled their advertising from the program.[28]

Drama is an important vehicle for addressing public conflicts and concerns. Even though it's impossible to be thoroughly objective in such presentations, producers can strive for balance and fairness to opposing points of view. Otherwise, issue-oriented drama turns into propaganda in favor of one perspective.

The miniseries *Amerika,* which depicted a Soviet takeover of the United States, was severely criticized by some people on the political left who believed that the show fostered a Cold War mentality and encouraged fascist sentiments.[29] As in the case of *The Day After,* the critics may have been wrong in their assessment of the program's point of view. Nevertheless, more balance in both *Amerika* and *The Day After* would probably have encouraged greater public discussion and less heated rhetoric.

In the 1980s, dramas about sexual abuse (*I Know My Name Is Steven*[30]) and wife battering (*The Burning Bed*[31]) have given thousands of people the courage to seek help or to intervene in cases of abuse. Since these are indeed important social issues, it is well within the scope of television to address them on behalf of the nation—as long as they are treated with fairness and sensitivity.

Too often Christians assume that public issues are only matters of personal concern. In fact, even some of the most seemingly private activities are part of the very fabric of society. The "uneasy conscience" of American evangelicalism still drives the church toward personal piety at the expense of public morality.[32]

Conclusion
Morality is inescapably part of the world of television. As the major storyteller of our time, the tube addresses morality in the process of entertaining viewers.

Christians have an important role to play as fair-minded guardians of public morality, caretakers of their own homes and church communities. But Christians must be wise viewers and critics of televisual morality. Otherwise they will be dismissed as self-serving bigots or pig-headed moralists.

First, we must avoid knee-jerk moralism. The Christian community sometimes reacts before it has the correct information or adequate time to reflect. It can embarrass all believers when any one of us criticizes the networks for airing programs that we have never even seen. Correct information is essential in making moral judgments.

Sometimes we'll have to view programs we would otherwise reject. Many parents will not let their children watch MTV even though they have never viewed it themselves. Yet adolescents know that not everything on the channel is morally evil. Misinformed parents, like reactionary viewers in general, can easily lose their credibility when they level unsubstantiated charges against television programs, cable channels and videotaped movies.[33]

Second, we need spiritual discernment to help us see beyond the big three moral issues—sex, violence and profanity. My own view is that materialism is by far the biggest moral problem with television. Dramatic shows, comedies, sports and commercials all preach worldly success and idolize consumption. As C. S. Lewis suggested in *The Screwtape Letters,* the biggest threat to Western Christians might be their own desires for popularity and material success, for all the hedonistic pleasures. Yet Christians continue to ignore such deep threats to the faith. Which is really a greater challenge to the church in America: middle-class comfort, what Lewis called a "bourgeois mind," or sexual promiscuity?[34] We should be willing at least to ask such tough questions for the sake of moral discernment. The overall point of view of most television programming is strongly materialistic.

Third, Christians need to realize that artistic quality cannot be reduced to morality. By focusing narrowly on morality, we have encouraged networks to offer aesthetically inferior programs. We should expect far more than inoffensive shows, and the Cultural Mandate demands it. Serious, issue-oriented drama is important for society, and we should seek it out on both commercial and public television. We have the right to expect balance in those shows, and to criticize the people who make purely sensationalistic programs. But well-done drama is a special gift that we should all cherish—even when it addresses sensitive moral questions.

Fourth, it is important for Christians to band together in television watchdog or advocacy groups. These organizations can help orchestrate our concerns as well as give Christians a unified voice to praise producers and networks that are providing morally and artistically superior programming.

As things now stand, however, these groups tend to be both moralistic and cleverly self-serving. They whip up public outrage partly as a means of raising funds for their organizations. Along the way, they often encourage people to write in protest to networks and sponsors without first viewing the offending program.

We need groups that will be as concerned with injecting public conscience and group solidarity into the public sphere as they are with getting the media spotlight on themselves and raising contributions. The best groups will avoid "rhetorical imprecision and myopic campaigning" in favor of a broadly defined concern with the public good.[35]

Finally, every family should openly establish its own moral standards for selecting programs. The problem of what to watch is really an opportunity for Christian families to build spiritual discernment in the home. Young children have to be patiently nurtured to understand the kinds of moral criteria provided in this chapter. If adults do not set the example by their own viewing selections, standards will not go very far. Families need to view and discuss programs together as they chart a moral course with the tube. Then children learn that the Christian life, like television viewing, is an ongoing search for the will of God in a world filled with both grace and sinfulness.

7
THE SOUL
OF HOLLYWOOD/
Who's Responsible for TV Fare

*Fallen! Fallen is Babylon the Great! . . . The merchants of the earth grew rich from her
excessive luxuries. (REV 18:2-3)*

I N HER KISS-AND-TELL MEMOIR TITLED *YOU'LL NEVER EAT LUNCH IN THIS
Town Again,* former movie producer Julia Phillips portrayed Hollywood as
a battle zone among ambitious but often untalented people. She summarized
her view of the entertainment industry this way: "Hollywood is a place that
attracts people with massive holes in their souls."[1]

Hollywood society can be very unforgiving, especially to members who
publicly criticize others. Phillips undoubtedly sold a few books with her salvos, but she spoke her mind only after Hollywood had already written her off.
Her revealing autobiography made money, but it could never get her back in
the business.

The television business—part of the massive entertainment industry known
as Hollywood—is governed by ambitious but often terribly insecure people.
The major movers and shakers number only in the hundreds. They make the
rules for tens of thousands of employees.

Thousands more—hopeful writers, actors, directors and producers—are living in Los Angeles and elsewhere, waiting for a chance to make it in "the business." If they ever get their chance, these hopeful souls will soon realize that the business can be far worse than frustrating. Hollywood's rules are slippery and pragmatic, but newcomers have to learn to play by them. If they do not, their success could soon be over.

The commercial television industry is like a social club, but not always a pleasant one. As one friend, a successful producer, puts it, "Hollywood is like living in junior high school in hell." The result is often emotional chaos and dysfunctional families—like the ones in Ron Howard's film *Parenthood*. Ben Stein says that there are more than two thousand Alcoholics Anonymous meetings in Los Angeles every week, and many are "jammed with people in the entertainment business."[2]

Anyone who seeks membership had better learn the rules of the game. Public expressions of personal agendas are frowned upon. The most successful members sometimes take on the values and beliefs of the industry while shedding their own "souls," to use Phillips' word. Like a street prostitute who will do almost anything to turn a profitable trick, some members of the industry will do whatever the market demands. They have no self—no sense of who they are apart from their desire to succeed. They become hired artists who help tell somebody else's tale, not their own.

At the same time, however, some individuals with integrity discover that the market wants what they happen to be talented enough to produce. This tension between venal artistry and personal integrity reflects the uneasy soul of the entertainment industry.

Christian television viewers should understand how that tension shapes the programs they watch. Critical viewing requires a fairly sophisticated perspective on the attitudes, values and beliefs that motivate people in the industry. Indeed, the social institution of commercial television is nothing more than the rules and behaviors of the business.

Ron Powers believes that television is not about entertainment as much as it is about "corporate arrogation and corporate will."[3] Public, commercial and religious programming are different because the people who make them are usually committed to widely different visions of what shows to produce, why to produce them, whom to produce them for and how to judge their quality. The rules of the game shape their own sense of the value and purpose of culture.

To put it in biblical perspective, the television industry marches to the orders of conflicting kings, all of whom provide their own cultural mandate. Producers and viewers need to be able to assess the allegiances behind these mandates as a way of evaluating their own televisual stewardship.

The issue of this chapter, then, is crucial to redeeming television. It's not enough for us to criticize or applaud televisual productions. We must acknowledge and evaluate the minds and hearts behind the programming.

Television is more than a business of entertainment. It's also a cultural activity carried out by people who are part of a larger industrial community. Edward Carnell wrote decades ago that those "who control television are responsible to all the people in the world, not to a limited group in any single country."[4] Perhaps this is even more true today, in the age of international television. TV's creators are workers in fields of culture, harvesting their own productions to the glory of someone—or, more typically, some institution, such as a production company or a television network.

Television Conspiracies

Christians have long held to various conspiracy theories to explain the products of Hollywood. These types of theories flourish in climates of ignorance and fear. Early movies, which often addressed sexual themes and portrayed sensuous love affairs, caused many believers to wonder whether Hollywood was bent on corrupting the nation's morals. Today a number of Christian organizations and individuals still advance conspiracy theories to explain the values and beliefs apparent in movies.

Television elicited similar conspiratorial concerns in its early years, but never as pervasively as film. Probably because television is so popular, and is viewed in the home instead of at a theater, it has not been as prone to such attacks. Viewers mistakenly think they know a lot about television simply because they watch it. The tube is perceived as a more intimate medium, closer to the real lives of viewers and less the product of a distant, evil empire.

The truth is that the movie and television businesses are increasingly the same, because of new technologies such as cable and the VCR as well as because of media mergers and buyouts. Just about any criticism that can be leveled at the people who create television programs can be applied to those who work in the movies—and vice versa.

In any case, there are still conspiracy theories about television, especially among the more conservative, fundamentalist communities. But the industry is also criticized by both the political right and the left, as well as by non-Christians. Although the most popular view is that television is owned and operated by liberals, there are conspiracy theories that take the opposite point of view. These irreconcilable theories are ways for people to explain a rather complex process: how and why particular programs end up on the air.

All such theories assume that the television business is composed of individuals who secretly choose programming with their own selfish interests in mind. Paradoxically, the many conflicting theories are all correct about this point, even though they come to widely divergent conclusions. There are many kinds of selfishness, from the love of money to the drive for celebrity status or corporate power. Souls can have many different kinds of holes.

Television programming *is* the result of various conspiracies. There are conspiracies between the networks and production companies, among the staffs of each television program, between actors and producers, and so on. But each of these conspiracies is little more than the natural result of television production and, in the case of commercial broadcasting, the free-enterprise system.

Television, unlike painting or sculpture, is inherently a collaborative endeavor that requires many people to work together in order to make a show, sell it to a station or network, promote it to an audience and especially to get the funding for the next one. But rarely are these conspiracies formed principally to harm people or to gain power over them. All the conspiracy theories fail when they attempt to explain the eventual products of collaborative work as the simple result of anyone's self-interest.

Tim LaHaye and Secular Humanism. One of the more vocal advocates of the theory that television programming is a liberal conspiracy is Tim LaHaye. In his 1980 book *The Battle for the Mind,* LaHaye cited "the media" as one of the four "vehicles of mind control." "When television licenses became available," he wrote, "humanists flooded the field. . . . It is obvious, by the degenerate programming that has appeared in recent years, that the three major networks (ABC, NBC, and CBS) are predominantly controlled by amoral humanists. A medium that once featured family-oriented programming and observed discretionary moral standards now makes jokes about homosexuality, incest, wife swapping, and depravity."[5] In 1984 LaHaye elaborated his conspiracy theory in *The Hidden Censors,* which linked the American media to the Council on Foreign Relations and the Trilateral Commission, both of which have been cited in the past by the paranoid John Birch Society.[6]

LaHaye's views seem to be supported by some surveys of television production executives. The movers and shakers in the industry are generally liberal and do seem to be more humanistically inclined toward secularism. One study of 104 executives found that 75 per cent describe themselves as "left of center politically," compared with only 14 per cent who place themselves right of center. Four out of five of them voted for George McGovern for president in 1972, while only one in five voted for the popular Ronald Reagan in 1980. Similarly, 75 per cent of the surveyed television executives thought the government should reduce the income gap between the rich and the poor, and 44 per cent said that government should guarantee employment to anyone who wants a job.

The most remarkable findings, however, were the executives' liberal attitudes toward social and moral issues. Nine out of ten of them strongly agreed that a woman has a right to choose abortion. Over half the respondents indicated that adultery is not wrong, while almost three-quarters said the same for homosexuality. In spite of the fact that 93 per cent of the executives had a "religious upbringing" (59 per cent Jewish), 45 per cent claimed no religious affiliation, and 93 per cent said they seldom or never attend religious services.[7]

Nevertheless, such survey findings do not necessarily validate the theories of conservative critics such as LaHaye. First, such surveys are probably unrepresentative of the business. Certainly a survey of executives would not reflect the values and beliefs of all people in the television industry. Second, even if the executives are as liberal as the surveys suggest, the programming itself does not necessarily reflect their own views of society and personal morality. After all, the commercial television industry is in the business of attracting the largest possible audience to sell to advertisers. Audience ratings reveal that sensational and sexually explicit programming tends to attract audiences.

Conservative conspiracy theories wrongly assume that the primary *intent* of television workers is to shape the values and beliefs of Americans. There is considerable contrary evidence that liberal programming is more the product of the social institution than it is a mere reflection of personal views of people in the industry.

The industry teaches newcomers how to write and produce, how to direct, how to schedule TV programs. In other words, the rules of the business shape how workers think and act. The TV business demands considerable allegiance to the industry's existing values and beliefs, which are the result of hard-nosed experience in a highly competitive, market-driven environment. Anyone who enters the business intent on pursuing his or her personal agenda is not likely to succeed—unless that agenda is profitable for the industry. Political or moral conspiracies don't make the business tick. Money does.

James Hitchcock and Pseudo-Sophisticated Culture. James Hitchcock offers a conspiratorial view of television programming that better takes into account the institutional values of the commercial industry. In *What Is Secular Humanism?* Hitchcock sees the media conspiracy as a natural outgrowth of the way commercial television is financed, not a covert political movement. He argues that advertisers want to reach the more liberal, upwardly mobile social groups with spendable income. This lucrative audience is "eager for 'sophisticated' entertainment," is "open to all points of view" and is "anxious to see 'controversial' subjects explored 'frankly.' These are people who must eventually shatter all taboos because they deny themselves nothing."[8] Hitchcock further argues that "media people" themselves are part of this liberal social group, and that the "moral revolution" on the tube reflects their personal values as well.

There is considerable merit in Hitchcock's perspective. For one thing, there are many examples of programming that addresses controversial issues: "Thirtysomething," "Hill Street Blues," "L.A. Law." Audiences of such programs tended to be upscale. Certainly the shows' plots and characters displayed a sense of tolerance and personal moral relativism. For another thing, these types of shows tend to depict a culture that is more "sophisticated" than that of most programs. Part of their audience lure is a pseudo-sophistication that appears to be genuine, an image-making that is not highly conscious of itself. As a result, these types of programs offered a greater personal relevance to their viewers, who became emotionally attached to characters by identifying with their professional aspirations and personal dilemmas. Advertisers, in turn, increasingly realized the significance of that identification for peddling their products to upwardly mobile viewers.

However, Hitchcock's theory goes only so far. Even into the early 1990s, as cable television and VCRs whittled away at broadcast networks' audience share, the major networks still looked primarily for mass-audience programs like "The Cosby Show," unquestionably the family-oriented blockbuster of the 1980s. Even programs such as "Full House" and "Family Matters" could not be explained by Hitchcock's theory, not to mention older shows such as "Little House on the Prairie" and "The Waltons," which were quite popular in their day. Hitchcock's theory helps explain some, but not all, of the trends in programming.

Ben Stein and the View from Sunset Boulevard. Writer and attorney Ben Stein took a very different approach to the subject of television's conspiracy. He interviewed television writers, producers, performers and directors, meeting them in country clubs, restaurants, offices and homes. In short, he immersed himself in the Hollywood culture. His conclusion after a year was simple but provocative: "Television is spewing out the messages of a few writers and producers (literally in the low hundreds), almost all of whom live in Los Angeles. Television is not necessarily a mirror of anything besides what those few people think. The entertainment component of television is dominated by men and women who have a unified, idiosyncratic view of life."[9]

Stein's immersion into Hollywood society clarified its country-club lifestyles and relative insularity from middle America. Few Americans can afford million-dollar homes, hundred-dollar lunches, Jaguar automobiles and all the rest. But his analysis of the relationship between those lifestyles and the resulting programs was tenuous. We can joke about the fact that sitcom families have large kitchens, but that is more the result of the need for an adequate-sized set than it is a reflection of Southern California lifestyles. Similarly, the tube's portrayals of crime, the clergy, the poor, the wealthy and government are not mere reflections of the makers' views.

In an insightful collection of interviews with television producers, Horace Newcomb and Robert Alley found some evidence to support Stein's theory. These television executives, among the most powerful in the industry, did sometimes consciously shape their programming after their own values and beliefs—though not always in ways that reflected a Hollywood culture. Earl Hamner modeled "The Waltons" after his own family experiences in the South. Norman Lear admitted that his series, from "All in the Family" to "Maude," sometimes expressed his own "thoughts" and "attitudes" and those of his collaborators: "We all had children, we all had marriages, we all had concerns as citizens of the world and of this country. Our humor expressed some of those concerns."[10]

Garry Marshall, whose productions included "Laverne and Shirley" and "Mork and Mindy," said that he tried to make society's negative images more positive. The moral of his shows was "Be nice to each other. Don't hate."[11]

But Newcomb and Alley also got the producers to discuss how their creativity was limited by the demands of the marketplace or by artistic considerations. James L. Brooks recalled how "The Mary Tyler Moore Show" was not meant to advocate women's rights but to establish a fictional world where such rights were being talked about, as they were in society at the time.[12]

Quinn Martin, who created "The F.B.I.," considered himself "much more politically left" than the show's portrayal of the agency. He said, "Whether I was liberal, conservative, moderate, whatever—you can have an overall point of view, but you shouldn't use your vehicles to try to do a polemic. . . . I don't think that's what our business is."[13] And producers Richard Levinson and William Link recalled developing the character Columbo "for purposes of drama and juxtaposition and contrast, not to advance their own Hollywood life styles."[14]

Stein's view of Sunset Boulevard is also only part of the story. It helps explain the content of some programs, but not others. Like LaHaye and Hitchcock, Stein wants to turn a partial truth into the whole truth. Perhaps most damaging to Stein's theory is that many television executives simply did not grow up in Hollywood. Their views of life are broader than Sunset Boulevard.

Todd Gitlin and Prime-Time Uncertainty. On the opposite end of the political spectrum from Tim LaHaye is scholar Todd Gitlin, whose book *Inside Prime Time* is one of the most extensive portraits of the workings of the television business. Like Stein, Gitlin spent considerable time interviewing television executives. His conclusion is surprising. Although he is a leftist critic of the media, Gitlin concluded that the element that most shapes the commercial television industry is not politics or moral values, but sheer uncertainty.

No one in the industry knows exactly which shows will be popular and which will flop in the audience ratings. Moreover, there seems to be no sure way to predict success; the business is mired in its own chaos. Executives can do little but follow a kind of televisual folklore: female leads rarely are a hit, half-hour shows are more successful than hour-long dramas, comedy attracts more viewers than serious drama, and so on.

Gitlin believes that the internal organizational values of the television industry determine programming choices. Those values are fundamentally monetary—the largest audience means the greatest advertising revenues. However, since no one knows for sure what programs will attract the largest number of viewers, corporate capitalism invariably leads to uncertainty in the business. The result is not right-wing programs, per se, but "conservative" programming designed in the hope of luring audiences. When programs challenge audiences' values, it is usually because of the industry's miscalculations, not because of some moral or political conspiracy.

Gitlin's conclusions are probably the most sensible and supportable of the "conspiracy" theories, but they do not explain all TV programming. Gitlin's thesis works best for explaining new series that are imitations of existing popular ones. It fails to explain why the networks would take many risks on the innovative programs of a writer-producer like Stephen Bochco ("L.A. Law," "Doogie Howser," "Cop Rock"), even though his "Hill Street Blues" was a significant success.

Gitlin has cogently shown how crazy the television business really is because of the unpredictability of success. The industry is incapable of a full-fledged political or moral conspiracy. Instead, the business "conspires" to keep itself afloat financially by chasing its own tail of past achievements. The heavy weight of profitability is an enormous albatross around the industry's neck.

A Producer's Medium

Producers shape commercial television programs—especially the major writer-producers. They lead the rest of the writing staff and oversee the making of the entire series. Their job is to orchestrate the storytelling, even when they don't write the original script. Compared with feature film writers and producers, they have enormous creative clout.

As Richard Levinson and William Link once put it, the television producer is "the engine that drives a series."[15] Or, as Russian-born producer David Victor ("The Man from U.N.C.L.E." and "Marcus Welby, M.D.") said, the producer is the person with the "vision or concept" for a show.[16] He or she exercises more control over the complex efforts of collaborative production than any other person. "He (they *are* mostly males) must hire writers and develop scripts, cast the actors for every episode, select the directors, worry about shooting schedules and budgets, keep his cast and crew happy, edit, score, and dub each segment, and see to it that the series receives its fair share of on-air and print promotion."[17]

Consider the remarkable cases of two producers, Hamner and Lear. Hamner not only conceived the idea that led to "The Waltons" but also wrote the show's pilot, "The Homecoming," oversaw the series' many changes in story, casting and design, and even produced a documentary study of the show near its final episode. If all of that were not enough, Hamner narrated the program, serving as the voice of a mature John Boy who recalls the historical context that opened and closed each episode of the show.

During Lear's most successful period in the late 1970s, his weekly duties as executive producer of a number of comedies included these:

Mondays meant reading "All in the Family" and a run-through for "Maude," "One Day at a Time," and "The Jeffersons." That meant a camera run-through and two back-to-back tapings for each. . . . Wednesdays, we read "Maude," "One Day at a Time," and "The Jeffersons," had a rehearsal hall run-through for "All in the Family" and a stage run-through for "Good Times." On Thursday, we taped "Good Times" and had a run-through on "All in the Family." Fridays, we read "Good Times," had a rehearsal hall run-through on "Maude," "One Day at a Time," and "The Jefferson," and taped "All in the Family." I also did most of the warm-ups for the audience before the tapings.[18]

Not all writer-producers are as busy as Hamner and Lear, but most of these creative visionaries are overworked individuals battling chaos. Producers are part artist, part entrepreneur, part personnel manager, part promoter and part organizer. It's impossible for them to oversee all the details of creating and selling television series, but they are ultimately in charge. If there is a single mind or heart behind a series, it is usually theirs. They are the closest thing to identifiable makers of television drama, but the viewing public knows little or nothing about them.[19]

As Stein recognized, they are part of a small club. In the fall of 1990 there were about 150 executive producers for all the prime-time series on ABC, CBS, Fox and NBC. Most of them had produced shows previously. Veteran producers are almost always preferred to neophytes, even when the veterans' early shows were flops. They develop series within the limits of what the networks or stations will buy from them.

Few producers can personally bankroll their own programs; they are at the mercy of corporations, primarily studios, for their financial blood and creative authority. Among the large studios are Columbia Pictures Television, MCA (Universal) Television, Paramount Television, Twentieth Television and Warner Brothers Television. These companies alone provided capital for forty-five of the series aired in the fall of 1990.

Most television producers are big-time operators in a bigger world of national and even international finance and entertainment. As television scholar David Marc aptly put it, television has become "the art of business." [20]

Large program budgets and corporate control have not always been good for television. Ironically, some television drama is not as good today, when an hour-long series episode can cost a million dollars to produce, as it was in the early years of the tube known as "the Golden Age." Even critic Neil Postman admits this. He suggests that some of the anthology drama of the late 1940s and early 1950s included some of the finest moments on the tube. Those teleplays began as adaptations of classical and contemporary fiction, but soon included primarily original works written specifically for the new medium. Among the more gifted screenwriters were Horton Foote, Gore Vidal, Rod Serling and Paddy Chayefsky, who later wrote the sardonic film about television called *Network*.[21]

In those days the tube provided special opportunities for gifted writers who were not producers. Consequently, the medium was shaped less by business considerations and more by literary and artistic ones—as is contemporary public television, another writer's medium.

The Real Censors
In commercial television, producers' freedom is always limited by the market for their programs. Creative control does not guarantee a market for their work. They have to sell their shows to someone.

Producers can sell their shows directly to individual television stations (a system known as syndication). But the major market for drama (as opposed to game shows, tabloid TV programs and the like) is the networks, including ABC, CBS, NBC and Fox, and more and more the cable networks. In other words, even though producers are the creative souls of commercial television, they work indirectly for stations and networks, their financial lifeblood.

For good and for bad, networks largely decide what shows will get on the air. They are the real censors who distribute particular programs to viewers while keeping others off the air and cable. Only after they select which programs to purchase and broadcast do the audience ratings have an opportunity to be implicit censors—and even then only with network consent.[22]

One of stations' and networks' most important considerations is whether a program will fit with scheduling plans. They decide whether to purchase the rights to air a program based partly on "audience flow"—the attempt to capture and hold viewers. To them, programs are like pieces of a jigsaw puzzle. Every half-hour period—especially during prime time, 8:00 to 11:00 P.M. (Eastern Standard Time, Monday through Sunday)—is part of the puzzle.

The major techniques of scheduling programs to maximize audience flow include the following:

1. *counterprogramming:* scheduling the most popular programs concurrently with a competitor's least successful shows

2. *block programming:* scheduling programs with similar audience appeal (for example, family sitcoms) adjacent to each other

3. *strong lead-ins:* scheduling high-rated shows at the beginning of prime time

4. *creating a hammock:* scheduling poorly-rated programs between two higher-rated ones

5. *stunting:* keeping the competition off-balance by making rapid schedule changes, beginning new series with longer episodes and the like[23]

Few programs are aired simply because of their own merits. Instead, they are broadcast because they will presumably help establish the best possible audience flow. Remote-control sets and cable television have made this ever more difficult for broadcast networks, who now have to compete for audiences with dozens of other programmers. Since the advent of cable and the VCR, average broadcast-network audiences have declined from about 90 per cent of viewers to around 60 per cent.[24]

As viewers get more fickle, and faster at the remote trigger, networks are increasingly in a quandary about how best to schedule their shows. Networks would like every show they run to lure a large audience, when in fact every year the whole process of scheduling programs becomes more tenuous. Most new shows every season are canceled before spring, as each network tinkers competitively with its schedule in hope of finding the right line-up of programs.

Network scheduling is one form of network censorship. Of course the networks are concerned about how a particular program will be received by the audience. They don't want to offend viewers unduly for fear that some people might change channels. But the network program schedulers know that a bit of titillation or sensationalism can attract viewers as well. Often bad publicity is better than no publicity. After the Fox network was publicly criticized for the sexual references in "Married . . . with Children," audience ratings increased very significantly. Apparently viewers wanted to hear the sexual language for themselves.

Network censorship is driven by pragmatic business considerations, even though it's not always clear what would be the best business decision. Issues of morality, artistic quality and public interest are nearly always secondary to audience ratings and advertising revenues. Ethical considerations are collapsed into marketing judgments.

The recurring justification is predictable: Networks and stations merely give viewers what they want to watch. Unfortunately, because of viewers' fickleness and the vicissitudes of the program marketplace, that justification cannot be proven or disproven. All we can say for certain is what the audience ratings were on a particular night—not what they would have been if the network censors had made different decisions or if viewers were more discerning.

The Syndication Gamble

At the heart of the television industry is a remarkable willingness to gamble megadollars on particular series. As strange as it seems, production companies normally sell their shows to networks at a considerable deficit. Networks usually pay the production company only part of the cost of making the shows. With every drama the company sells to the networks, it loses more money. The production company undertakes this gamble in the hope that eventually a program will be so popular that the production company can make money in rerun syndication.

In the fall of 1990, for example, only twelve series were sold to networks for what they cost to produce. The rest of the network programs were sold for deficits estimated at between $25,000 and $400,000 per episode. New World Entertainment produced each episode of "The Wonder Years" for about $750,000, but received only $600,000 from the ABC network. Similarly, the company that made "Golden Girls" lost an estimated $50,000 on every episode. The program with the largest deficit that season ("The Flash," at

$400,000 per episode) was canceled by CBS before the end of the year.[25]

These incredible losses are acceptable in the television production business because of the enormous profits that are sometimes earned by selling reruns to individual stations—a process known as rerun syndication. It costs production companies to deal with the networks, but it also helps guarantee the profitability of reruns.

In 1983, for example, "Hill Street Blues" reruns were sold to four major television stations for an estimated $30 million. Each of the stations was permitted to run episodes of the series up to ten times during an eight-year period beginning in 1987. The company that produced "M*A*S*H" grossed through syndication an estimated $1 million from each half-hour episode— a total of more than $255 million.[26]

Bill Cosby, who shared in the syndication rights of his popular comedy program, in the early 1990s became the king of moneymakers in the entertainment industry, according to *Forbes*. By 1992 he made an estimated $180 million on syndication alone. A few years earlier, *Time* had appropriately called him "Cosby, Inc." on the magazine's cover.[27]

Whenever a program is a network success, it is a likely candidate for mega-revenues from rerun syndication. The higher the network audience ratings, the greater the eventual syndication windfall. And the more series a company produces, the better its chances of eventually hitting the jackpot. So on and on go the producers, cranking the networks' arms, like slot machines, in hopes of getting their programs on the air. With as much luck as skill, a producer will eventually make a hit series for the syndication market. Few businesses are similarly predicated on making consumer products at a loss while waiting for a big payout to bankroll new ones.

Risk pervades the making and scheduling of prime-time drama. It is not the calculated risk of a well-researched product. Television is much closer to the speculative risk of the commodities markets, where brokers hedge their bets against a backdrop of incalculable uncertainties.

Over the long run, there are fewer winners cloaked in syndication royalties than there are losers still playing the game. Even the unsuccessful television producer is hardly a bohemian artist; losers are paid well for trying—often hundreds of thousands of dollars per year. But the real winners gain their wealth from the syndication payoff: one television gambler hits the unpredictable jackpot while another one stumbles.

Conclusion

Like all artists and businesspeople, the makers of television drama are motivated by many different things—from the love of their craft to a desire for status and wealth. As an institution, commercial television encourages mixed motives as long as they help produce network hits and syndication bonanzas. The bottom line is profit, not quality or integrity. Of course it's possible for some producers to be concerned about all three, but dollars speak louder than the authenticity of one's work or the craftsmanship of one's creations. It takes particularly talented producers to sell high-quality shows to networks or stations.

Our responsibility as Christians is to help redeem the institutions of television, including the organizations that make and distribute programs. If television as a technology belongs to the Lord, so too do the social institutions of the medium. Not everyone is called to be a television writer, editor, director, actor or producer, but all of us can influence the way society uses the technology, including the way the business of television operates.

First, Christians should be among the supporters of noncommercial television, especially public television. We may not like all of the programming on public television, but it provides additional choices over commercial television. In fact, as a writer's medium, public television reflects some of the heterogeneity of contemporary folk, popular and fine art. Public television's vision is not guided principally by audience ratings, so not every program strives for mass appeal.

Many of the people who work in public television are paid considerably less than their commercial counterparts. They are often in the medium for the more altruistic goal of community service. Along with that goal come greater institutional integrity and smaller "holes in the soul." Public television deserves our financial support.

Some readers will undoubtedly question the extent to which public television is Christian. Certainly a nonprofit "business" is not explicitly religious, let alone Christian. Moreover, the apparent world view of much public television is very secular. This is true of some of the documentaries and nature shows as well as the drama.

Nevertheless, day in and day out public television is considerably more moral and redemptive than commercial television, especially the four major commercial networks. Moreover, public television is much more concerned with artistic quality.

Second, Christians should communicate with commercial television stations, producers, networks and advertisers. Letters and telephone calls may not always produce satisfactory results, but the television business needs to hear more than the audience ratings. Our voices can be effective when our arguments are reasonable and articulate. We can be leaders in encouraging more than financial values in the commercial television business. It usually takes massive letter-writing campaigns to change network executives' minds about which programs to schedule, but it takes only a few letters to remind producers as well as networks and stations that they have a real audience, not just a passive, manipulable one.

Because the writers and producers are less well known than the actors, we wrongly tend to ignore them. Even a handful of letters to them can make a difference. They are not used to much feedback, especially positive letters. One of the best ways to encourage the better writers and producers is to send them letters of support when they are creating programs with redeeming value. Their names are given with every episode, and they can be reached through the network or production company, whose addresses can usually be secured through one telephone call to the program manager of the local station.

By all means pray for producers who are making quality programs. They are in a tough business and need our support.

Third, the Christian community should encourage talented young people to enter the television business. In the long run, the most effective way to transform the industry is to bring into it gifted individuals who do not have massive holes in their souls. This is the type of conspiracy the medium needs: an infusion of talented people with personal integrity, authentic stories to tell and a moral conscience.

Too many talented people have been discouraged under the erroneous assumption that television is unredeemable. Movie actress-director Jodie Foster credits her "business morality" and sense of responsibility for her success—even more than her talent.[28] Not everyone is called to redeem the tube, but surely some Christians are. Yet without the encouragement of other believers, they may not heed the call.

Because the tube is so dependent on storytellers, Christians should have an edge in the business. The church is one of the few institutions in society that values narratives. Only the entertainment industry competes formidably with the church on the basis of who can tell better tales. As ethnic and racial communities lose their historic stories, they will not be able to challenge the tube's stories very effectively.

Christians are a people of the book of historic tales—stories of faith and hope, tragedy and despair. They should know the pathos of life, and they should be able to tell others about life through compelling tales. Not surprisingly, some of the most successful television and film writers are from religious backgrounds.

Perhaps the best that can be said of the television business is that there is enough uncertainty in it to let in the visionary ideas of some individuals without holes in their souls. If business formulas were cut and dried, if ratings could be predicted, if syndication revenues were not a gamble, it would be a very different industry. The old rules of the game can be cast aside when someone comes up with new ones that work in the marketplace. Given the right people in the business, through the grace of God television could be a much more redemptive enterprise.

8
REDEEMING
THE TUBE/
What Each of Us Can Do to Improve TV

Faith by itself, if it is not accompanied by action, is dead. (JAS 2:17)

*T*HE EARTH AND EVERYTHING IN IT BELONGS TO GOD, SAYS THE PSALMIST.
Humankind has been entrusted with the Lord's creation. We—all of us—are
to be responsible stewards of God's world. Television is now part of that
world, for good and bad. How we use the tube will determine whether or not
God is praised and this part of the creation is cared for and developed.

Earlier we saw that many people adopt one of two misguided approaches
to using the medium. First, many critics assume a gnostic view of television
as inherently evil. Instead of focusing on the potential of the tube, they become
preoccupied with its alleged negative effects. From intellectuals to common
folk, the gnostic critics, whether or not they are Christians, see salvation from
the tube only in its elimination. They have little sense of the sovereignty of
God and the goodness of the creation.

Vocal proponents of television charge off in the opposite direction. Whether
televangelists or videophiles, they are bearers of technological optimism. Such
advocates of television focus on the alleged positive effects of the tube—

including religious conversions. Along the way they overlook the fallenness of human beings and the distortion of God's creation. They too easily link television with the will of the Lord or the altruistic motives of humankind.

The problem of television, then, echoes the existential dilemmas faced by all God's children: how to steer a wise course between naive condemnation and absurd idolatry, between the goodness of the creation and the sinfulness of humankind.[1] Previous chapters took steps toward such wisdom, but the goal of redeeming television requires still more discernment and critical Christian scrutiny.

This final chapter establishes an overall mandate for Christians who want to help redeem the tube to the glory of God. It provides a fivefold path to follow: (1) discriminating viewing, (2) visual literacy in schools and churches, (3) greater ethicality in commercial television, (4) alternatives to commercial television and (5) better-informed television criticism.

1. Every Christian should be a discriminating viewer who wisely uses the medium of television.

We must begin with our own lives. It's easy to criticize the television industry, but far more difficult for most of us to get our own viewing in order. Like the overall population, most American Christians are lazy, selfish viewers who say one thing and practice another. As inheritors of the Fall, we seek immediate gratification and personal pleasure rather than the kingdom of God.

Holiness requires Christians to select their television programs carefully so as not to be conformed to this world. The tube reflects the disobedience and rebelliousness of human beings in contemporary culture. Indeed, the tube tends to accept the world the way it is and make it even more that way. To put it differently, the medium typically confirms what people in this fallen culture *want* to believe—for example, that sexual attractiveness is the basis of personal worth, that material prosperity is a worthwhile goal in life, that evil comes only from evil people, not from all of humankind.[2] Discriminating viewing requires us to see programs for what they really are in the light of the gospel, not in the shadow of our own tastes and desires.

Undiscriminating adults tend to pass along the same attitude to their own children. I am repeatedly astounded at the kinds of videos and programs that parents permit their offspring to see.[3] Even as toddlers, many children are plunged into televisual stories with adult themes, graphic violence and sick humor. Apparently many parents simply do not believe that what their kids view will leave any lasting imprint on their minds and hearts. How wrong they are! No one knows all of the effects, but there is no question that the tube contributes role models for early school-age children and identities for teenagers.

The biblical view of communication links it inextricably with culture and identity. Communication is the way culture is passed from person to person, from society to the individual and from generation to generation. Every time there is communication among people, they are changed, however slightly, into the image of someone or something.

Sitting uncritically in front of the tube for hours on end, people are gradually shaped into the image of the people on the set. They are less and less holy, because it is ever more difficult to distinguish them from nonbelievers and the secular culture. Their overall attitude toward life is nurtured by television producers and programmers, not by the church, pastors, parents, teachers or friends. They "commune" via the television with the stars and celebrities they have come to admire—the prelates of prime time. And they often believe what these personas tell them to be true, even if they reject the particular messages of a given show.

Every family should have its own standards for viewing. Holiness requires the family to know what it stands for and what programs are therefore worth watching. Restrictions on viewing are not enough. Too many parents try to take the easy way toward holiness by naively and ignorantly rejecting culture.

Christian parents must nurture children to be discerning about the surrounding culture, including television. Young people deserve viewing standards that link restrictions to reasons for not viewing some programs and for viewing other ones.

Reasons, not just restrictions, are the televisual currency of genuine discernment. If rock video shows or evening soap operas are off-limits, children ought to be told why. And if parents are unable to give reasons, they should set about the task of getting adequate information to make such judgments. Ignorance too often leads to a kind of blindness that masquerades as great righteousness.

Gross televised immorality can be discerned quickly, but what about the more subtle forms such as racism, sexism and ethnocentrism? Often Christian parents are quick to point out their righteous viewing standards, which are actually little more than restrictions on the obvious depiction of sex, violence and profanity. Children deserve more cultural discernment.

It seems to me that every family should encourage its members to enjoy the full richness of the medium. Discriminating viewing looks for worthwhile programs, not just worthless ones. And there *are* worthwhile shows across the spectrum of storytelling. There are excellent programs to amuse, instruct, confirm and illuminate. Those who sample only one kind of fare miss out on much that the tube has to offer by the grace of God.

Probably the biggest mistake that parents make along these lines is creating a home environment that uses television only as a vehicle for mindless entertainment. Most of the college students I have taught over the years have never viewed a single serious drama on public television; they have no idea that such fare even exists. This is truly a lamentable situation for adolescents from Christian homes and privileged backgrounds. Their parents apparently protected them, but from the wrong programs. The "safety" of popular shows led to little or no appreciation of television as a dramatic medium. The Cultural Mandate cannot be fulfilled on television exclusively by situation comedies, soap operas, detective shows and the like, no matter how morally inoffensive they are.

Moreover, discriminating viewing requires us to make conscious choices among all the activities that compete for our limited time and energy. Anyone who fills discretionary time with the tube is a blessing to the television industry—but not to others.

Every decision to watch the tube is also a decision not to do something else, whether spending time with our loved ones, serving the church, enjoying nature or praying for the needs of the world.

The tube is often like an enormous vacuum that sucks up people's God-given talents by cutting them off from other people. It can be an opiate that makes viewers dependent on programs and personalities, from the news to late-night movies. Just as the drug addict can become a burden rather than a contributor to society, the nondiscriminating television viewer can become a weight on the shoulders of family, friends, coworkers and church. In this sense many families are televisually dysfunctional.

Every time the set is turned on, the Christian viewer and his or her family ought to ask this fundamental question: Is this what God would have me do with my time and talent? Sometimes the answer may be "yes," even for the sake of amusement, but certainly not as often as is decided implicitly in many homes every night.

The new television technologies offer little help for nondiscriminating viewers. Cable television and the VCR, for instance, do not automatically make Christian viewers more discerning. To the contrary, they expand program options and further test viewers' capacity to make tough decisions. New technologies require *more* knowledge and *greater* wisdom on the part of all viewers, but especially parents, who must contend with a seemingly never-ending barrage of more options.

Without fairly clear standards, most families are lost in the sea of possibilities. There is little to do, then, except to let every individual watch whatever he or she pleases. Then individualism triumphs over family or community, technology over society and relativism over standards. That is why in some homes there are three, four and even five television sets—one for each individual.

2. Christians should push for televisual literacy in schools and churches.

Christians, especially evangelicals, are often among the first to use new mass-media technologies. In the name of evangelism, they seize each medium to reach more people in less time. However, Christians are also usually naive about the impact of the technology on the message and the people being evangelized. Too often action comes before understanding, and the long-term implications do not surface for years or decades.

The Protestant emphasis during the Reformation on *sola scriptura* (Scripture only) helped liberate the church from some nonbiblical traditions, but it also created new problems. The emphasis on the printed word over oral tradition eventually created an individualistic approach to reading the Bible in which everyone began to interpret Scripture for him- or herself. One result is the huge number of contemporary denominations.[4]

Because every medium communicates its own way, and has its own advantages and disadvantages, it is imperative that people understand the media they use. As I argued earlier, no medium is neutral with respect to its effects on both messages and the people who use it. As a technology and a social institution, television massages—to use Marshall McLuhan's term—the individuals and society that employ it.[5] There is no way around this. Like ripples in a pond, communications media radiate intended and "accidental" effects. Unless we understand the implications for the tube, we will be doomed to the kind of world that television establishes.

Schools and churches have both failed to educate people about television, even though it is the dominant mass medium of our time. Schools are ill-equipped to teach students how to interpret and evaluate televisual messages. Similarly, many local churches and denominations take pot shots at TV instead of articulately addressing the ways the tube communicates. Ironically, mainline Protestant churches have done far more than evangelical groups, who are thought by many to be champions of television because of their dominance in religious broadcasting. The National Council of Churches and the United Church of Christ have contributed more to serious public policy discussions about television than have all of the evangelical churches combined. Yet they, too, have failed to provide overall educational direction to local churches and schools.[6]

Today the schools are so widely criticized that there is little merit to adding one more voice of discontent. Nevertheless, I think that my concern is sufficiently different from most criticism that it deserves special consideration. Certainly the schools can do more to improve learning in traditional subjects, what has been called "back-to-basics." And anything that interferes with that movement, including televisual education, is potentially damaging to the nation. However, television is so ubiquitous and influential a public medium that it cannot be ignored by the schools.

Even Neil Postman, perhaps the best-known critic of the medium, believes that schools *must* assume the central role in teaching children how to be critical viewers.[7] The schools are society's principal institution for educating young people to be informed, intelligent and rational citizens. And since television is now part of the very fabric of public life, as well as the most influential storyteller, it deserves to be on the national educational agenda.

In my opinion, the worst way to introduce television into the educational system is as a completely independent subject of study. Because television is a medium, not a discrete subject area, it should be integrated into many subjects across the curriculum. Just as students learn to read critically about history or the sciences, they should be taught to use televisual resources (films and videotapes) as part of the educational process.

Once they see that televisual communication is *about* something, and not just entertainment, students are on the road to becoming critical viewers. At first most young students are completely taken aback by this idea, largely because they have been socialized to consider the tube a source of relief from boredom, not a communications medium. Eventually, however, students can be taught to interpret and evaluate what they see and hear on television, just as they can be taught to interpret the written word.

In public schools it can be difficult to get beyond the common concerns of public life, such as civic virtue and individual freedom. Because of shared beliefs among teachers, students and parents, Christian schools can bring spiritual discernment directly to bear on the issue of critical viewing. Unfortunately, limited funds and small staff often prohibit these schools from making much headway on the issue. In some communities the problem of limited resources is compounded by conservative parents who do not believe that Christian schools should address worldly concerns like television. Meanwhile, the students from these schools are soaking up hours of television every day—often more than their public-school counterparts—because parents believe the home is spiritually safer than the neighborhood.

Churches need their own back-to-basics education because of enormous biblical illiteracy among children and adults alike. Yet churches might find that one of the most effective ways of introducing the faith to young believers is a sustained examination of contemporary culture, including television. A critique of television and film, for example, can highlight both what Christians should believe and what the surrounding culture teaches or implies every day on television and at the movie theaters.

It is one thing to lead a Bible study aimed at uncovering the meaning of a scriptural text. It is far more to explicate the significance of the Scriptures for contemporary culture and society. Why not institute church-education classes to discuss currently popular television programs and films? Should pastors not address popular art in their sermons? Television provides an enormous pool of shared public icons, values and motives just waiting for biblical examination.

However schools and churches address the issue of television, they must keep two things in mind. First, any overbearing, one-sided attack on the medium will attract only people who already share that opinion. If schools and churches want to contribute fruitfully to people's televisual education, they will have to seek some kind of balance between good and bad, creation and fall, sin and grace. There is no sense preaching about television's evils to those who already believe that TV is evil. Moreover, this belief is simply wrong unless it is leavened with some appreciation for what the medium can accomplish with quality, integrity and morality.

Second, schools and churches must learn to pull the family together as a viewing unit. Because the tube is viewed primarily in the home, and because the family is the most powerful socializing agent, televisual literacy must be promoted ultimately by the family if there are to be long-term changes in people's viewing habits and standards.

Neither the church nor the school alone is the answer to establishing a televisually literate public, but they can foster stronger family commitments for the task. Television "turn-off" weeks and months can be encouraged by schools or churches, but they will not succeed without the cooperation of parents. Fathers, in particular, are roadblocks to family involvement in such activities. I've repeatedly witnessed how hard it is for some wives to gain their husbands' support in changing family viewing habits and even attending public presentations on the topic. Many men wrongly feel that their families "owe" them unlimited viewing time because of their hard work each day at the office or factory.

Progressive educators and pastors ought to think creatively about ways of using television to build literate students and discerning congregants. After all, television has to be redeemed _by_ someone; the medium will not redeem itself. We can hope for a wise viewing public and discerning parishes that are articulate critics of the tube but also appreciate good and worthwhile programming. Some of them would also be prepared to help transform the television business itself.

3. Christians should help redeem the institutions of commercial television.

Not just the technology, but the organizations that make and distribute television programs belong to God. This includes the major networks—CBS, NBC, ABC and Fox. It includes the production studios and independent producers. Today it also includes the growing number of video rental shops, the cable TV companies and the satellite networks distributed via cable. In short, the commercial institutions of television are a vast array of organizations that make money in one way or another on the tube.

In my judgment, the impulse in some Christian circles to completely abandon commercial television is misguided. As discussed in the last chapter, there is much about the business of television that merits criticism, and even condemnation. But commercial television remains a redeemable institution; the sporadic broadcast of truly entertaining programs, from series to movies, proves the point. So do some of the news, documentary, variety and special programs.

Commercial television is not entirely without conscience or hope, although it has frequently offered fare that is outrageously bad—morally and artistically. There are recurring signs of moral life, if not artistic integrity, on commercial television. These signs should be encouraged, and one of the best ways to do that is to thank networks, producers and stations for their attempts at worthwhile shows.

Orchestrated letter-writing campaigns have not been very effective at permanently changing the minds of broadcast executives, but they remain a potentially viable means of influencing these organizations. Too often Christians write to networks or advertisers only with complaints. Also, many conservative Christians are apt to write only on the basis of what other people, especially professional media watchdog groups, have told them about programs. The letters they send are typically moralistic tirades that focus on sex, violence and profanity while ignoring other aspects of programs, including the broader moral issues discussed earlier. Conducted in this manner, letter-writing campaigns will have little impact, because executives and advertisers easily stereotype the senders as nay-saying, uninformed, moralistic zealots who wish to impose their view of reality on everyone else.

I believe that public protests—including protests against television programming—are an important part of democratic life. But when such protests are orchestrated by self-interested watchdog groups partly for the purpose of raising funds, they lose their credibility.

Let's all protest as we see fit. And let's do it boldly and unashamedly. But let's not be deceived by semiprofessional media critics who crank out slick junk mail about how television is going to the devil and how their organization is going to save Hollywood. If we are not personally well informed, we ought not to write letters of compliment or condemnation. Integrity must undergird our actions. Perhaps the best long-term action we could undertake on behalf of better commercial television would be to encourage talented Christians to enter the industry. Not everyone is called to do this—nor should they feel as if they have to do it if they are not called by God.

There are already too many untalented Christians who want to change the industry with little more than a zeal for evangelism. Bible-thumping Christians will not get far in Hollywood—not because of what they believe, but because of how they act and because of their typical lack of ability as writers, directors and agents. A desire to save Hollywood personalities is not sufficient for redeeming television as an institution.

On the other hand, some Christians have the talent but do not have the spiritual maturity and strong support group necessary to survive the television industry with their faith intact. Talent and faith are essential for those who truly want to make a difference.

Our responsibility, if we ourselves are not called to work in television, is to identify and support people who are. Once again, American individualism is evident in the way the Christian community rarely offers such support. It's simply too easy to lay the burden of a calling completely on the individual being called by God. Sometimes the person may have a financial need, for schooling or transportation or temporary housing. Other times he or she will need wise counsel, prayer, fellowship or words of encouragement. I believe popular entertainment is so influential in our society that we should be ready to provide such support to talented and spiritually wise people called by God.

Years ago in many Christian traditions the primary calling was to the ministry—and rightly so, for ministers were among the most influential leaders in society. With the secularization of society, this has changed enormously; today ministry is not nearly so esteemed, even within the body of believers.

In one sense this is unfortunate, because the church desperately needs gifted preachers, pastors and evangelists. In another sense, however, secularization has opened new opportunities for lay ministry in and out of the church. The Reformation idea of the priesthood of all believers has exciting implications for all forms of work, including televisual endeavors. It is, biblically speaking, no less a calling to the entertainment industry than to agriculture, medicine and the ministry.

The Lord rules over television, as he does over the pastorate. Let's pray that he will raise up a generation of talented artists and businesspeople who can be caretakers of God's creation by making a difference in television.

The television industry is not the easiest business for college graduates to enter. During the 1980s and early 1990s, the broadcast networks were forced through competition with cable and the VCR, as well as through an economic recession, to lay off many talented people. Only the Fox network grew. At the same time, however, the expansion of cable and the booming syndication market created all kinds of new jobs.

Today the film and television production businesses are in flux, so that entry-level positions are open to many new people every year. Moreover, there are many untalented people in the industry who are able to hang onto their jobs only because they have experience. Persistent and talented individuals *can* get into the business. This is one of the best-kept secrets of Hollywood.

Finally, Christians can help redeem commercial television by expressing their support for any federal legislation designed to help ensure that the industry truly operates in the "public interest, convenience and necessity." The official regulating agencies, such as the Federal Communications Commission in the United States, are important bodies that supposedly represent the public. In order to do that well they must not get too wrapped up in politics, yet they must hear from the public they are mandated to protect.

When important business is before these organizations, we should send articulate letters to express our views. We can also inform our federal representatives, who can bring about actual changes in policy. My point here is not to advocate particular policy, but to encourage citizens to be informed and help shape the direction of television regulation.

4. Christians should help establish and maintain alternatives to commercial television.
Commercial television will never meet all the needs of society and the Christian community for quality television. There will always be worthwhile types of programming that can't be supported only through the sponsorship of advertisers. Moreover, there will invariably be explicitly religious programs that no advertisers will support; the producers will have to make hard decisions about whether or not to appeal directly to audiences for donations, thereby running the risk of distorting the show's gospel message in order to gain contributors.[8] Given these considerations, Christians would be well advised to establish and maintain viable alternatives to commercial television.

Currently there are two major alternatives to commercial television: public television and so-called Christian TV. Both deserve careful scrutiny and, in many cases, Christians' time and money. Unfortunately, both have serious drawbacks. Public television came of age in the 1960s as "educational television." Funded principally by the U.S. Congress, it was originally designed to

176 _____ *Redeeming Television*

fill an educational gap left open by commercial television, which was geared far more to entertainment than to instruction. By the 1970s programs such as "Sesame Street" and "Mr. Rogers' Neighborhood" had established loyal audiences of middle-class children. Yet few people were aware that public television was offering many types of excellent dramatic productions, a few series, outstanding musical concerts and some informative news and documentary shows.

Public television's major problem, aside from funding, was publicity. Too few Americans knew about public stations' programs. Nevertheless, public television became an increasingly important alternative to commercial broadcasting, which sometimes supported public television as an important community resource.[9]

In general, Christians should be generous supporters of public television, which increasingly needs local contributions of time and talent in order to survive. Not all of its programming will please all viewers, but this is true of any broadcast institution. On a day-to-day basis, public television provides significantly more worthwhile programming than do commercial stations and networks. That's because public television is free from the tyranny of audience ratings; it does not have to strive for the largest audience for every program. Instead, it can provide programs for a wide range of cultural tastes and ethnic backgrounds. It can also instruct viewers without the fear that most viewers will switch to a more entertaining channel.

The biggest problem with public television, from a Christian perspective, is a pervasive secularism. Like commercial broadcasting, public television does not often address specifically religious themes or issues. And its nature and science programs are likely to assume the perspective of evolutionary naturalism. Nevertheless, this does not mean that public television is antireligious; in fact, some of its science programs establish a kind of mystery about the universe that leaves open the possibility of a Creator God.

Drama on public television is not generally concerned with matters of faith, although there are numerous exceptions. Compared with commercial television, however, the drama on public broadcasting is much more likely to address universal topics such as death, evil and love. In addition, the quality of scripts, level of performance and technical execution are all consistently better on public television.

In my view, public television is more redemptive than commercial fare because there is significantly more artistic quality and moral integrity. We may not agree with the point of view of a particular show on public television, but we can be more certain that the program represents someone's point of view rather than simply pandering to the marketplace. Moreover, in my experience public television at the local level has been quite open to programs that challenge secularism.

Perhaps the best way to view public television is as a forum for a wide range of perspectives. It is up to Christians to contribute ideas as well as time and talent to help public television accomplish this goal.

Christian television has tried many programming ideas over the last twenty years, but two problems have plagued it severely. First, Christian stations and networks have almost always offered inferior imitations of nonreligious shows. They have tried religious variety shows, talk shows, game shows, soap operas and the like. In nearly every case the result has been technically and creatively inferior.

Second, Christian television has never established sensible ways of funding itself. Too often the stations and networks have taken the easy way out: selling program time to practically any religious broadcaster who will sign a contract, regardless of the quality of the show, its theological integrity and its ability to draw non-Christian viewers. The result has been schedules of often embarrassing programs that appeal only to small groups of contributors. So far Christian television has been able to do little more than keep itself hobbling along, often as a second-rate imitation of commercial broadcasting.

Christian television has limited its reach and quality by an immediate focus on the Great Commission as the only basis for religious programming. Let me be clear: Missionary television deserves our support. But we should also support Christian broadcasters who have a broader notion of their calling grounded in the Cultural Mandate—because in the long run this may be significantly more evangelistic.

We should encourage these ministries to consider the quality of their productions. Religious zeal is not a sufficient basis for launching a Christian station, network or program. Local cable-access programming has been the worst offender. Too many well-intentioned Christians make local shows that are more like home-video productions than professional broadcasts.

We ought to encourage talented young people who want to work in Christian television. Some of them wrongly believe that Christian broadcasting might be a lesser calling than commercial television, so they give up on their dreams. Like commercial television, religious broadcasting needs gifted individuals with a concern for both quality and integrity. The televangelism scandals of the late 1980s proved this point well. Christian television is even more of a shoestring operation than public television, and it needs the support of viewers as well as the talents of directors, writers, producers and the like. This type of broadcasting should be a viable and attractive calling, not just a last resort for those who cannot make it in the "big time."

5. Christians should encourage general-interest and religious media to provide serious, informed television criticism.
Nothing is more essential to redeeming television than a knowledgeable and critical viewing public. As things now stand, the general public and the Christian community are both remarkably naive about the tube—how it communicates, how programs are financed, the ways televisual stories work in people's lives, the role of government regulation in cable and broadcast television. Viewers need help interpreting television programming and following what goes on in the industry. These are largely the tasks of reviewers and critics. We need more of them.

Most of us are too busy with other matters even to stay on top of the better new television shows and videos available. Imagine what a service it would be to viewers if there were newspaper and religious-magazine columns that highlighted the best television series and new video releases. Consider as well the benefits of regular articles that would report on the work of talented television artists known for their integrity. Interviews with successful producers and writers might greatly help viewers understand how and why particular programs get on the air. All of this would help demystify the medium for the average viewer, who may not have the time or educational background to make sense of it on his or her own.

Christian periodicals, in particular, should do far more television reporting and criticism. Clearly they need gifted writers who understand the medium. But they also need the courage to take up this task. For most Christian magazines it is simply easier, and less controversial, to ignore all but the most outrageous programming, which they readily condemn. These media should spend as much time informing readers about worthwhile productions as they do blasting the immoral fare that networks distribute purely in the name of profit. And they should do a lot more of both if their goal is to serve the Christian community.

Once again, it's important for Christians to support young people who feel called to be television critics, regardless of whether they wish to write for general-interest or religious publications. In fact, I believe the Christian critic could have considerable impact in either type of medium. The need is there, even in secular media, where few people want to be television critics or even general entertainment writers. Video and television criticism can be a noble line of work and an important calling.

Another possibility would be for local congregations and Christian schools to organize their own television criticism. Parents and children could share their program reviews and viewing standards in weekly or monthly newsletters. Such newsletters could be either original reviews with viewing suggestions or reprints of articles from other sources, used with permission.

At the same time, church and school libraries could purchase tapes of some of the better programs and films available. An increasing number of television programs are available on videotape, and the vast majority of worthwhile films are as well. Schools and churches can heighten community awareness and provide practical help through these types of projects.

Conclusion

Television may be the Trojan horse of Western civilization. We invited it into our homes—first the set, then cable and the VCR, then big-screen models with stereo and high definition. Like the wooden horse, the tube seems so benign; then its sounds and images engulf viewers, families and nations. In only forty years television has captured virtually all the discretionary time of millions of North Americans. Along the way the tube has subverted what many people claim to stand for: strong families, moral character and democratic values. No aspect of culture is left untouched, from religion to sex to politics. All of it is processed through the electronic eye.

It's so easy for Christians either to throw up their hands in despair or to go along with the masses of indifferent watchers. At one public presentation, a thoroughly disillusioned mother pleaded with my audience to get rid of their television sets. I understood her frustration. Haven't we all wondered whether that might be the best solution?

Several hours later I collected my thoughts during a brisk evening walk through my neighborhood. The same bluish glow emanated from nearly every home. I thought to myself that these neighbors were not going to heed the woman's advice. And maybe they shouldn't.

Sometimes it's hard to maintain a spirit of hope in a setting of despair—to remember and believe that because God is ultimately in charge, our actions can make a difference. As Edward Carnell put it in the early days of television, "T.V. has placed the warfare of the two cities (of God and man) in the parlor. . . . lest a premature pessimism overtake us . . . let us quickly counterbalance early fears with a realistic study of T.V.'s hopes."[10]

I encourage all readers to make a difference with the electronic Trojan horse in their midst. For most of us, the best plan is to "think small" by concentrating on how we use the tube in our own homes. For some, it may be time to take bigger action.

In this book I've issued a positive call for Christians to seize control of the television. My prayer is that the church of Jesus Christ will not be coopted by either gnostic or idolatrous impulses, but will faithfully seek discernment. Then television will truly be redeemed.

Notes

Chapter 1: The Great Commotion

[1]Malcolm Muggeridge, *Christ and the Media* (Grand Rapids: Eerdmans, 1977).

[2]Neil Postman, *Amusing Ourselves to Death: Public Discourse in the Age of Show Business* (New York: Viking, 1984).

[3]Ashley Montagu, "Television and the New Image of Man," in *The Human Dialogue,* edited by Floyd W. Matson and Ashley Montagu (New York: Free Press, 1967), 360.

[4]Ibid., 359.

[5]I have documented this in Quentin J. Schultze, *Televangelism and American Culture: The Business of Popular Religion* (Grand Rapids: Baker, 1991), chap. 7.

[6]Kenneth A. Myers, *All God's Children and Blue Suede Shoes: Christians and Popular Culture* (Westchester, Ill.: Crossway Books, 1989), 22.

[7]H. R. Rookmaaker, *Art Needs No Justification* (Downers Grove, Ill.: InterVarsity Press, 1978).

[8]Gene Veith, Jr., *The Gift of Art* (Downers Grove, Ill.: InterVarsity Press, 1983), 20-21.

[9]Quoted in Abraham Kuyper, *Lectures on Calvinism* (Grand Rapids: Eerdmans, 1931), 153.

[10]Albert M. Wolters, *Creation Regained: Biblical Basics for a Reformational Worldview* (Grand Rapids: Eerdmans, 1985); Henry R. Van Til, *The Calvinistic Concept of Culture* (Philadelphia: Presbyterian and Reformed, 1974); Nicholas Wolterstorff, *Until Justice and Peace Embrace* (Grand Rapids: Eerdmans, 1983).

[11]T. S. Eliot, *Notes towards the Definition of Culture* (New York: Harcourt, Brace, 1949), 29.

[12]Raymond Williams, *Communications,* rev. ed. (London: Chatto & Windus, 1966), 19.

[13]James W. Carey, ed., *Communication as Culture: Essays on Media and Society* (Boston: Unwin Hyman, 1989), 13-36; Raymond Williams, *Keywords: A Vocabulary of Culture and Society* (New York: Oxford University

Press, 1976), 62-63.

[14]Joshua Meyrowitz, *No Sense of Place: The Impact of Electronic Media on Social Behavior* (New York: Oxford University Press, 1985), 116.

[15]Carey, *Communication,* 201-31; Daniel J. Czitrom, *Media and the American Mind: From Morse to McLuhan* (Chapel Hill: University of North Carolina Press, 1982). It should be clear by now that I am not equating "culture" with "high art," "elite art" or even "the arts." In my view, all the arguments about the superiority of various arts over others (for example, painting over architecture and cinema) cloud the most important thing: all cultural activities, all ways of life, are under the lordship of Christ. This does not mean that all cultural activity is equal, or that every work of art is of the same value as another. Rather, it suggests that Christians need to redeem all of the arts, from the most popular to the most elite, from highbrow to lowbrow, from West to East. I believe there are worthwhile and worthless products in all artistic and cultural categories, and that the Christian life should be a balance among them.

For an introduction to arguments for and against various ways of categorizing culture, see Herbert J. Gans, *Popular Culture and High Culture* (New York: Basic Books, 1974). Probably the best short discussion of this topic I have ever read is Raymond Williams, "On High Culture and Popular Culture," *The New Republic,* November 23, 1974, 13-16.

[16]Quentin J. Schultze, "The Mythos of the Electronic Church," *Critical Studies in Mass Communication* 4 (September 1987), 245-61; "Keeping the Faith: American Evangelicals and the Media," in *American Evangelicals and the Mass Media,* edited by Quentin J. Schultze (Grand Rapids: Zondervan/ Academie, 1990), 23-46.

[17]Merrill R. Abbey, *Man, Media and the Message* (New York: Friendship, 1960), 114.

[18]Virginia Stem Owens, *The Total Image: Selling Jesus in the Modern Age* (Grand Rapids: Eerdmans, 1980); Jacques Ellul, *The Humiliation of the Word* (Grand Rapids: Eerdmans, 1985).

[19]Jerry Mander, *Four Arguments for the Elimination of Television* (New York: Quill, 1978).

[20]Ellul, *Humiliation,* 194.

[21]Muggeridge, *Christ,* 105-6.

[22]Owens, *Total Image,* 98.

[23]See Robert Cathcart, "Our Soap Opera Friends," in *Inter/Media: Interpersonal Communication in a Media World,* edited by Gary Gumpert and Robert Cathcart, 3d ed. (New York: Oxford University Press, 1986), 207-18.

[24]Schultze, *Televangelism,* 77-80.

[25]Owens, *Total Image,* 40.

[26]Muggeridge, *Christ,* 45.

[27]Ellul, *Humiliation,* 195.

[28]One of the most revealing examples of this is Alvin Kernan, *The Death of Literature* (New Haven, Conn.: Yale University Press, 1990). Kernan's contrast of "book culture" and "television culture" is so fraught with misinformation and rhetorical sleight-of-hand that one might well ask who would believe such nonsense. Who would agree that a book *inherently* tends "toward elaborate themes and form" whereas "the television spectacle is seen once in a flash"? Or that "story and plot" are *inherently* "minimal or nonexistent" in television (p. 150)? The answer, of course, is that some literary critics might believe these things. Ironically, Kernan also laments the fact that today too many books are published.

[29]Nathan Hatch, *The Democratization of American Christianity* (New Haven, Conn.: Yale University Press, 1989).

[30]One of the best books to address this issue is Gregor T. Goethals, *The Electronic Golden Calf: Images, Religion and the Making of Meaning* (Cambridge, Mass.: Cowley, 1990). Also see Tyron Inbody, ed., *Changing Channels: The Church and the Television Revolution* (Dayton, Ohio: Whaleprints, 1990).

[31]See Carey, *Communication,* 113-41.

[32]Schultze, *American Evangelicals,* 23-46.

[33]Jimmy Swaggart, "Divine Imperatives for Broadcast Ministry," *Religious Broadcasting,* November 1984, 14.

[34]See Schultze, *Televangelism,* 57-59.

[35]Ben Armstrong, *The Electric Church* (Nashville: Thomas Nelson, 1979), 8-9, 172-73.

[36]Willard D. Rowland, Jr., *The Politics of TV Violence: Policy Uses of Communication Research* (Beverly Hills, Calif.: Sage, 1983), 23.

[37]Examples of these arguments abound in the popular press. The last is articulated by Henry J. Perkinson, *Getting Better: Television and Moral Progress* (New Brunswick, N.J.: Transaction, 1991).

[38]One book that addresses the technological and social aspects of television is Peter Conrad, *Television: The Medium and Its Manners* (Boston: Routledge & Kegan Paul, 1982).

[39]The development of technology is not a "neutral" activity, but is embedded in particular values and purposes. Usually particular technologies are created, and others are not, because people have in mind some social purpose that steers the invention of technology. I don't have the space in this book to develop the argument, but it is developed with respect to television in Raymond Williams, *Television: Technology and Cultural Form* (London: Fontana, 1974).

[40]Stephen V. Monsma and the Calvin Center for Christian Scholarship, *Re-*

sponsible Technology: A Christian Perspective (Grand Rapids: Eerdmans, 1986), 19.

41Owens, *Total Image,* 41.

42Postman, *Amusing,* 160-61. For an analysis of Postman's changing position with respect to television and education, see Tommi Hoikkala, Ossi Rahkonen, Christoffer Tigerstedt and Jussi Tuormaa, "Wait a Minute, Mr. Postman! Some Critical Remarks on Neil Postman's Childhood Theory," *Acta Sociologica* 30 (1987), 87-99. Probably the best overall critique of Postman's views is found in Joli Jensen, *Redeeming Modernity: Contradictions in Media Criticism* (Newbury Park, Calif.: Sage, 1990).

43Tony M. Lentz, *Orality and Literacy in Hellenic Greece* (Carbondale, Ill.: Southern Illinois University Press, 1989), 7.

44Edward J. Carnell, *Television: Servant or Master?* (Grand Rapids: Eerdmans, 1950), 21.

Chapter 2: Tales on the Tube

1George Comstock, *The Evolution of American Television* (Newbury Park, Calif.: Sage, 1989), 44.

2Although this book emphasizes the American context, most of my ideas can be adapted to other countries and cultures. Moreover, since American programming is most popular throughout the world, and the U.S. is the largest exporter of entertainment, it makes good sense to concentrate on American fare. Readers interested in the Canadian context might want to supplement this book with these studies of Canadian television programming and regulation: Mary Jane Miller, *Turn Up the Contrast: CBC Television Drama since 1952* (Vancouver: University of British Columbia Press/CBC Enterprises, 1987); Paul Rutherford, *When Television Was Young: Primetime Canada 1952-1967* (Toronto: University of Toronto Press, 1990); S. M. Crean, *Who's Afraid of Canadian Culture?* (Don Mills, Ont.: General Publishing, 1976).

3David Marc, *Demographic Vistas: Television in American Culture* (Philadelphia: University of Pennsylvania Press, 1984), 1-6.

4Postman, *Amusing.*

5Steve Allen, *Dumbth and Eighty-One Other Things to Make Americans Smarter* (Buffalo, N.Y.: Prometheus, 1989).

6Postman, *Amusing,* 160.

7The value of "diversionary pleasure" on television was argued already in 1950 by evangelical theologian Edward J. Carnell in his excellent book *Television: Servant or Master?,* 29-33.

8Academic studies of television addiction are somewhat contradictory in their findings. It seems clear that there is sometimes such a thing as dependency, but whether or not this is biological is uncertain. Marie Winn's popular writings on this topic are not particularly well supported with research. See her *The Plug-in Drug* (New York: Viking, 1977).

[9]William F. Fore, *Image and Impact: How Man Comes Through in the Mass Media* (New York: Friendship, 1970), 108.

[10]"How TV Is Creating New Perceptions about Blacks," *The New York Times,* February 7, 1991, B1, B4.

[11]Andrea L. Press, *Women Watching Television: Gender, Class and Generation in the Television Viewing Experience* (Philadelphia: University of Pennsylvania Press, 1991), 175.

[12]Tamar Liebes and Elihu Katz, "Dallas and Genesis: Primordiality and Seriality in Popular Culture," in *Media, Myths and Narratives: Television and the Press,* edited by James W. Carey (Newbury Park, Calif.: Sage, 1988), 88-112; *The Export of Meaning: Cross-Cultural Readings of "Dallas"* (New York: Oxford University Press, 1990). For other views on cross-cultural impact see Christian W. Thomsen, *Cultural Transfer or Electronic Imperialism? The Impact of American Television Programs on European Television* (Heidelberg: Carl Winter Universitatsverlag, 1989), and Felipe Korzenny and Stella Ting-Toomey, eds., *Mass Media Effects across Cultures* (Newbury Park, Calif.: Sage, 1992).

[13]Sonia M. Livingstone, *Making Sense of Television: The Psychology of Audience Interpretation* (Oxford: Pergamon, 1990), 56-57.

[14]Joseph Turow, *Playing Doctor: Television, Storytelling and Medical Power* (New York: Oxford University Press, 1989), xv, 271.

[15]Letty Cottin Pogrebin, *Growing Up Free: Raising Your Child in the '80s* (New York: McGraw-Hill, 1980), 393.

[16]One study of the relationship between moral development and television viewing among youth is Kevin Ryan, "Television as a Moral Educator," in *Television as a Cultural Force,* edited by Richard Adler and Douglass Cater (New York: Praeger, 1976), 111-28. The toy issue is discussed splendidly in Tom Engelhardt, "The Shortcake Strategy," in *Watching Television,* edited by Todd Gitlin (New York: Pantheon, 1986), 68-110.

[17]National Center for Education Statistics, U.S. Department of Education, *A Profile of the American Eighth Grader* (Washington, D.C.: Government Printing Office, 1990), 48.

[18]Quentin J. Schultze et al., *Dancing in the Dark: Youth, Popular Culture and the Electronic Media* (Grand Rapids: Eerdmans, 1991).

[19]Schultze, *Dancing.*

[20]"Movies, TV Flood Teens with Sex," *Grand Rapids Press,* November 12, 1990, B3.

[21]The study was commissioned by Josh McDowell Ministry of Dallas in 1987, and was reported nationally by the Associated Press. "Conservative Church Teens Sexually Active, Researchers Discover," *Grand Rapids Press,* Febru-

ary 2, 1988, A3.

22Neil Postman et al., *Myths, Men and Beer: An Analysis of Beer Commercials on Broadcast Television, 1987* (Washington, D.C.: AAA Foundation for Traffic Safety, n.d.).

23Many thinkers have explored this idea; among them are Stanley Hauerwas, *A Community of Character* (Notre Dame, Ind.: Notre Dame University Press, 1981); Northrop Frye and Robert D. Denham, *Myth and Metaphor: Selected Essays, 1974-1988* (Charlottesville: University Press of Virginia, 1990); Nicholas Wolterstorff, *Works and Worlds of Art* (Oxford: Clarendon, 1980); Bruno Bettelheim, *The Uses of Enchantment* (New York: Knopf, 1976); Roland Barthes, *Mythologies,* translated by Anne Lavers (New York: Hill and Wang, 1972). See also volume 42, no. 2 (Summer 1990) of *Communication Studies* and vol. 34, no. 2 (1987) of *Media Development.*

24Ella Taylor, *Prime-Time Families: Television Culture in Postwar America* (Berkeley: University of California Press, 1989), 4.

25The mythic function of television has been examined in dozens of books and articles. Among them are several by William F. Fore: *Television and Religion: The Shaping of Faith, Values and Culture* (Minneapolis: Augsburg, 1987); *Mythmakers: Gospel, Culture and the Media* (New York: Friendship, 1990). Also see Michael R. Real, *Mass-Mediated Culture* (New York: Prentice-Hall, 1977); Gregor T. Goethals, *The TV Ritual: Worship at the Video Altar* (Boston: Beacon, 1981); Roger Silverstone, *The Message of Television: Myth and Narrative in Contemporary Culture* (London: Heinemann Educational Books, 1981); Robert Rutherford Smith, *Beyond the Wasteland: The Criticism of Broadcasting* (Urbana, Ill.: ERIC, 1976). David Thorburn calls this mythic function of stories "consensus narrative"; see his article "Television as an Aesthetic Medium," in Carey, *Communication,* 48-66.

26Jacques Ellul, *Propaganda* (New York: Knopf, 1971).

27One of the best surveys of TV genres is Brian G. Rose, ed., *TV Genres* (Westport, Conn.: Greenwood, 1985).

28Ruth Rosen, "Search for Yesterday," in Gitlin, *Watching Television,* 66.

29Stuart M. Kaminsky with Jeffrey H. Mahan, *American Television Genres* (Chicago: Nelson-Hall, 1985).

30David Marc, "Understanding Television," *The Atlantic,* August 1984, 34.

31Hal Himmelstein, *Television Myth and the American Mind* (New York: Praeger, 1984), 5.

32"Lay Pipe, Add Heat, Get Laughs," *Harper's Magazine,* November 1988, 45-55.

33Mark Crispin Miller, *Boxed In: The Culture of TV* (Evanston, Ill.: Northwestern University Press, 1988), 74.

34An interesting but somewhat dated study of stereotyping is Randall M. Miller, ed., *Ethnic Images in American Film and Television* (Philadelphia:

Balch Institute, 1978).

[35]Wolterstorff, *Works and Worlds*, 144.

[36]"Through the Eyes of Djeli Baba Sissoko: The Malian Oral Traditions," *The Speech Communication Teacher* 5 (Winter 1991), 1-2.

[37]Victor Lielz, "Television and Moral Order in a Secular Age," in *Interpreting Television: Current Research Perspectives*, edited by Willard D. Rowland and Bruce Watkins (Beverly Hills, Calif.: Sage, 1984), 284-85. Also see Horace Newcomb, *TV: The Most Popular Art* (Garden City, New York: Doubleday, 1974).

[38]I develop these beliefs more fully in "Secular Television as Popular Religion," in *Religious Television: Controversies and Conclusions*, edited by Robert Abelman and Stewart M. Hoover (Norwood, N.J.: Ablex, 1990), 239-48; and "Television Drama as a Sacred Text," in *Channels of Belief: Religion and American Commercial Television*, edited by John P. Ferré (Ames: Iowa State University Press, 1990), 3-28. John Wiley Nelson says that "the most consistent principle" of the American belief system is that "the source of evil is always external to those suffering the effects of the evil." See his *Your God Is Alive and Well and Appearing in Popular Culture* (Philadelphia: Westminster, 1976), 34.

[39]J. Fred MacDonald, *Who Shot the Sheriff? The Rise and Fall of the Television Western* (New York: Praeger, 1987), 135. Also see Quentin Schultze, *Television: Manna from Hollywood?* (Grand Rapids: Zondervan, 1986), 82-98.

[40]I have taken this up at more length in "Secular Television," 239-48; "Television Drama," 3-28; and *Television*.

[41]Julie D'Acci, "The Case of Cagney and Lacey," in *Boxed In: Women and Television*, edited by Helen Baehr and Gillian Dyer (New York: Pandora, 1987), 203-25.

[42]Ronald Berman, *How Television Sees Its Audience: A Look at the Looking Glass* (Newbury Park, Calif.: Sage, 1987), 9-10.

[43]Nelson, *Your God*.

[44]Nelson, *Your God*, 26.

[45]The debate over *The Day After* is summarized in Kathryn C. Montgomery, *Target Prime Time: Advocacy Groups and the Struggle over Entertainment Television* (New York: Oxford University Press, 1989), 4-5.

[46]Carnell, *Television*, 113.

[47]C. S. Lewis argued that most art is not meant to comment upon life, in spite of what some scholars seem to believe. In his view whether human artifacts comment upon life is not relevant in the question of whether or not the artifacts are art. He worried that such narrow thinking reduced art to philosophy: All art is *addition* to life. See C. S. Lewis, *An Experiment in Criticism* (Cambridge: Cambridge University Press, 1961), 81. Lewis's view

helps correct any overemphasis on illumination, but it hardly solves the question of what art is.

48Wolterstorff, *Works and Worlds,* 146.
49Tony Schwarz, *Media, the Second God* (New York: Random House, 1981), 163.
50See Schultze, *Televangelism,* 125-52.
51Also see Richard V. Peace, "The New Media Environment: Evangelism in a Visually-Oriented Society," *Journal of the Academy for Evangelism in Theological Education* 1 (1985-86), 36-45.
52"Americans and the Arts: VCR's Take Off," *Newsweek,* March 28, 1988, 69.

Chapter 3: Grazing Videots

1Amy Stuart Wells, "Video School Bus Enhances Education, Halts Disciplinary Problems," *Grand Rapids Press,* March 31, 1991, B2.
2Bruce Edwards, Jr., "The Medium and the Mediator: Viewing TV Christianly," *Mission Journal,* May 1981, 6.
3Solomon Simonson, *Crisis in Television* (New York: Living Books, 1966), 163.
4Mary Kane, "Researchers Say We Have More Free Time Than Ever," *Grand Rapids Press,* October 16, 1991, D1.
5"Average Daily Viewing in November," *Electronics Media,* February 10, 1992, 20.
6Margaret Morse, "An Ontology of Everyday Distraction: The Freeway, the Mall and Television," in *Logics of Television: Essays in Cultural Criticism,* edited by Patricia Mellencamp (Bloomington: Indiana University Press, 1990), 193-221.
7Michael Arlen, *The Camera Age* (New York: Farrar Straus Giroux, 1976), 142.
8Patricia Leigh Brown, "Where to Put the TV Set? There Are Still No Easy Answers," *The New York Times,* October 4, 1990, B4; "Where the TV Sets Are," *Electronic Media,* November 5, 1990, 50.
9*America's Watching, 1959-1989* (New York: Television Information Office, 1989), 8.
10Edward Rothstein, "Is Home Where the Set Is?" *The New York Times,* January 3, 1991, B2.
11Carnell, *Television,* 118-20.
12"When Bigger Isn't Better," *Channels,* June/July 1984, 10. Also see "The New TV Viewer," *Channels,* September 1988, 53-62.
13"Comedy on Television: A Dialogue," in *Television: The Creative Experience,* edited by A. William Bleum and Roger Manvell (New York: Hastings House, 1967), 97.
14Selling "spots" on television, rather than program sponsorship, was called

the "magazine" concept. See Erik Barnouw, *Tube of Plenty: The Evolution of American Television* (New York: Oxford University Press, 1975), 190.

[15]Douglas Brode, "The Made-for-TV Movie: Emergence of an Art Form," *Television Quarterly* 18 (Fall 1981), 78.

[16]Richard Levinson and William Link, *Stay Tuned* (New York: St. Martin's Press, 1981), 154.

[17]Ellen Goodman, "Are We Couch Potato Voters?" *Grand Rapids Press,* March 10, 1988, D4.

[18]"Hours of Cable TV Usage per Week by Households," *Electronic Media,* November 19, 1990, 41.

[19]Perry M. Smith, *How CNN Fought the War: A View from the Inside* (New York: Birch Lane, 1991), 132.

[20]James Barron, "To the Complaint 'Nothing on TV,' 150 Remedies," *The New York Times,* December 19, 1991, B1, B5.

[21]"The Magic Is Gone," *Grand Rapids Press,* October 11, 1990, D1.

[22]Quoted in Fred Schruers, "Peter Pandemonium," *Premiere,* December 1991, 74.

[23]Jerzy Kosinski, *Being There* (New York: Bantam, 1970). This book was later made into a film starring Peter Sellers.

[24]Mark Crispin Miller, "Deride and Conquer," in Gitlin, *Watching Television,* 228.

[25]Greg Paeth, "Randy Michaels Upsets WLW Audience," *Cincinnati Post,* May 31, 1990, 8C.

Chapter 4: Televisual Literacy

[1]"Bill Moyers," *American Film,* June 1990, 44; *Weekly Guide,* January 11, 1988, 1; Rushworth M. Kidder, "TV Made Hearings a Show and Oliver North a Star," *Grand Rapids Press,* July 28, 1987, A14.

[2]How TV images have influenced political discourse is discussed cogently in Kathleen Hall Jamieson, *Eloquence in an Electronic Age* (New York: Oxford University Press, 1988).

[3]Meyrowitz, *No Sense,* 98, 101.

[4]Postman, *Amusing,* 76-80; Neil Postman, *The Disappearance of Childhood* (New York: Delacorte, 1982), 114.

[5]David Riesman, with Nathan Glazer and Reuel Denney, *The Lonely Crowd* (Garden City, N.Y.: Doubleday, 1954).

[6]Ellul, *Humiliation,* 184.

[7]Ellul, *Humiliation,* 190-91.

[8]Goethals, *Electronic Golden.*

[9]Joe Holland, "Our Electronic Culture: A Theological Analysis," unpublished paper, Warwick Institute, 1991, 8.

[10]Michael Novak, "Toward Television Criticism," *Commonweal,* April 11,

1975, 40.

[11]Jean Shepherd, *A Fistful of Fig Newtons* (New York: Doubleday, 1981).

[12]For example, see what is probably the most influential study of the "language" of television from a structuralist perspective: John Fiske and John Hartley, *Reading Television* (London: Methuen, 1978).

[13]"The TV Producer: A Dialogue," in Bluem and Manvell, *Television,* 23.

[14]Guy Lyon Playfair, *The Evil Eye: The Unacceptable Face of Television* (London: Jonathan Cape, 1990), 14.

[15]See Schultze et al., *Dancing,* chap. 7.

[16]For an introduction to these cinematic devices, see Harold M. Foster, *The New Literacy: The Language of Film and Television* (Urbana, Ill.: National Council of Teachers of English, 1979), 4-12.

[17]"TV Producer," 23.

[18]Martin Esslin, *The Age of Television* (San Francisco: W. H. Freeman, 1982), 28.

[19]Jan Scott and Charles Lisanby, "Design for Television," in Bluem and Manvell, *Television,* 287-89.

[20]R. Zoglin, "Cool Cops, Hot Show," *Time,* September 16, 1985, 62.

[21]Esslin, *Age of Television,* 29.

[22]Donald Horton and R. Richard Wohl, "Mass Communication and Parasocial Interaction: Observations on Intimacy at a Distance," *Psychiatry* 19 (1956), 215-29.

[23]One of the best essays on this process is Jimmie L. Reeves, "Television Stardom: A Ritual of Social Typification and Individuation," in Carey, *Media, Myths,* 146-60.

[24]Marshall McLuhan, *Understanding Media* (New York: McGraw-Hill, 1964).

[25]Esslin, *Age of Television,* 30.

[26]Georgia Dullea, "When Celebrities Divorce, Good Will Has Top Billing," *The New York Times,* April 15, 1988, 17.

[27]James Endrst, "Audience Nostalgia Can Smother TV Stars," *Grand Rapids Press,* December 22, 1988, D8.

[28]Quoted in *Time,* May 9, 1988, 89.

[29]Sarah Ruth Kozloff, "Narrative Theory and Television," in *Channels of Discourse: Television and Contemporary Criticism,* edited by Robert C. Allen (London: Methuen, 1987), 44-45.

[30]Richard Schickel, *Intimate Strangers: The Culture of Celebrity* (Garden City, N. Y.: Doubleday, 1985), 11.

[31]Esslin, *Age of Television,* 22.

[32]Esslin, *Age of Television,* 28.

[33]Critical viewing could be taught in religious education programs. For essays on how religious educators might approach this, see John L. Elias, "Religious Education in a Television Culture," *Religious Education* 76 (March/

April 1981), 195-203; Pamela Mitchel, "Approaches to Television in Religious Education," in Inbody, *Changing Channels,* 97-112.

Chapter 5: TV Criticism

[1] William Mahoney, " 'Northern Exposure' Wins Poll," *Electronic Media,* November 18, 1991, 36, 38.

[2] "Season-to-Date Ratings/Shares of Prime-Time Programs for September 16 to December 15," *Electronic Media,* December 23, 1991, 19.

[3] Richard Zoglin, "Goodbye to the Mass Audience," *Time,* November 19 1990, 122.

[4] Ron Powers, *The Beast, the Eunuch and the Glass-Eyed Child: Television in the '80s* (New York: Harcourt Brace Jovanovich, 1990), 213-19.

[5] I hesitate to give the names of reviewers who might fall into this category, so I have concentrated on providing information about the better critics. There is a third category of critics that I can't figure out what to do with. Prominent among them is Marvin Kitman, whose ego-involved approach to criticism is interesting to read and sometimes quite insightful. However, it's usually more about Kitman and his travails than it is helpful evaluation or interpretation. See Marvin Kitman, *I Am a VCR* (New York: Random House, 1988).

[6] Anyone interested in the subject of television criticism and the critics should see the following: Miller, *Boxed In;* Thorburn, "Television as an Aesthetic Medium"; Tania Modleski, ed., *Studies in Entertainment* (Bloomington: Indiana University Press, 1986); Mary Ann Watson, *Critical Studies in Mass Communication* 2 (March 1985), 66-75; Les Brown, "Remarks to the Iowa Critics Conference," *Critical Studies in Mass Communication* 2 (December 1985), 390-95; Horace Newcomb, "American Television Criticism, 1970-1985," *Critical Studies in Mass Communication* 3 (June 1986), 217-28; Kathryn Montgomery, "Writing about Television in the Popular Press," *Critical Studies in Mass Communication* 2 (March 1985), 75-79; Patrick D. Hazard, "TV Criticism: A Prehistory," *Television Quarterly* 2 (Fall 1963), 52-60; Hal Himmelstein, *On the Small Screen* (New York: Praeger, 1981); Robert Sklar, *Prime-Time America: Life on and behind the Television Screen* (New York: Oxford University Press, 1980); VandeBerg and Wenner, *Television Criticism.*

[7] David Littlejohn, "Thoughts on Television Criticism," in Adler and Cater, *Television as a Cultural Force,* 148-49.

[8] See Erica Franklin, "Tubby Tubers," *American Health,* November, 1989, 88 and Ellen Stark, "Couch Potato Physique," September, 1989, 8.

[9] Rowland, *Politics of TV Violence.*

[10] Todd Gitlin, *Inside Prime Time* (New York: Pantheon, 1983), 273-324.

[11] Eileen R. Meehan, "Why We Don't Count: The Commodity Audience," in

Mellencamp, *Logics of Television,* 117-37.

[12]N. D. Batra, *The Hour of Television: Critical Approaches* (Metuchen, N.J.: Scarecrow, 1987), 243.

[13]William Henry III won the Pulitzer Prize for his work at the *Boston Globe.* He now writes for *Time and Channels.* See Michael J. Arlen, *Living Room War* (New York: Penguin, 1966); Powers, *The Beast, the Eunuch;* Harlan Ellison, *The Glass Teat* (New York: Ace Books, 1973); Les Brown, *The Business behind the Box* (New York: Harcourt Brace Jovanovich, 1971).

[14]Batra, *Hour of Television,* 218.

[15]Horace M. Newcomb, "One Night of Prime Time: An Analysis of Television's Multiple Voices," in Carey, *Media, Myths,* 88-112.

[16]One of the best collections of readings on the historical interpretation and explanation of television is John E. O'Connor, *American History, American Television: Interpreting the Video Past* (New York: Frederick Ungar, 1983).

[17]Horace Newcomb and Robert S. Alley, *The Producer's Medium* (New York: Oxford University Press, 1983), 164.

[18]Thomas D. Cook, *"Sesame Street" Revisited* (New York: Russell Sage Foundation, 1975); Gerald S. Lesser, *Children and Television: Lessons from "Sesame Street"* (New York: Random House, 1974).

[19]C. S. Lewis once distinguished between "realism of presentation" and "realism of content." The former is "the art of bringing something close to us, making it palpable and vivid, by sharply observed or imagined detail"; the latter is fiction that is "true to life." He said that the two are independent, and he equated neither with masterpieces—they are matters of style. See C. S. Lewis, *An Experiment in Criticism* (Cambridge: Cambridge University Press, 1961), 57-59. Certainly television has far more of the latter type of realism. In this book "realism" is used to refer to content, not presentation.

[20]See Quentin J. Schultze, "The Private Lives of Private Eyes," *Christianity Today,* September 4, 1987, 71, and Leah R. Ekdom Vande Berg, "Dramedy: 'Moonlighting' as an Emergent Generic Hybrid," in Vande Berg and Wenner, *Television Criticism,* 87-111.

[21]See John J. O'Conner, "Retro-Mania: When TV Was for Families," *The New York Times,* November 22, 1991, B5.

[22]Scott Williams, "Fox Broadcasting Co. Pursues Its TV Audience," *Grand Rapids Press,* November 23, 1991, C5. Also see "Tops with Teens: 'Beverly Hills,' " *USA Today,* October 31, 1991, D3.

[23]Gitlin, *Inside Prime Time,* 63-85.

[24]William Boddy has pointed out that critics felt in the 1950s that the shift from live performance and teleplay to the recorded series "signaled a retreat by the industry from an earlier commitment to aesthetic experimentation, program balance and free expression." But those "golden" years were never so golden as many believed. Each era seems to have quality television of its

own sort, from the teleplays of the 1950s to the social sitcoms of the 1970s. And there is always much forgettable fare as well. See William Boddy, *Fifties Television: The Industry and Its Critics* (Urbana: University of Illinois Press, 1990).

[25]Quoted in Irv Broughton, *Producers on Producing: The Making of Film and Television* (Jefferson, N.C.: McFarland, 1986), 193.

[26]Schultze et al., *Dancing.*

[27]H. R. Rookmaaker, *Modern Art and the Death of a Culture* (Downers Grove, Ill.: InterVarsity Press, 1970), 228-29.

[28]Veith, *Gift of Art,* 85.

[29]Martin Marty, "We Need More Religion in Our Sitcoms," *TV Guide,* December 24, 1983, 2-8.

[30]Clifford G. Christians, "Redemptive Media as the Evangelical's Cultural Task," in Schultze, *American Evangelicals,* 331-56; Clifford G. Christians, "Redemptive Popular Art: Television and the Cultural Mandate," *Reformed Journal,* August 1980, 14-19.

[31]William F. Fore, *Image and Impact: How Man Comes Through in the Mass Media* (New York: Friendship, 1970), 31.

[32]Wolterstorff, *Until Justice and Peace Embrace.* Also see his *Art in Action* (Grand Rapids: Eerdmans, 1980).

[33]Wayne C. Booth, "The Company We Keep: Self-Making in Imaginative Art, Old and New," *Daedalus* 3 (Fall 1982), 57. For a more extensive look at Booth's views on this subject, see his *The Company We Keep: An Ethics of Fiction* (Berkeley: University of California Press, 1988).

Chapter 6: Beyond Moralism

[1]Ken Auletta, *Three Blind Mice: How the TV Networks Lost Their Way* (New York: Random House), 364.

[2]Bill Carter, "NBC Defends Retreat from 'Quantum Leap' Episode," *The New York Times,* October 1, 1991, B2; Thomas Tyrer, " 'Quantum' Fight Sparks Debate," *Electronic Media,* October 7, 1991, 1, 31.

[3]Cal Thomas, "A Vulgar Society Shouldn't Be Surprised," *Grand Rapids Press,* October 16, 1991, A12.

[4]*Electronic Media,* September 30, 1991, 13.

[5]"Porn King Outstrips Others in Slow Economy," *Grand Rapids Press,* October 21, 1991, B3.

[6]See Donald E. Wildmon with Randall Nulton, *Donald Wildmon: The Man the Networks Love to Hate* (Wilmore, Ky.: Bristol Books, 1989).

[7]Quoted in John Brady, *The Craft of the Screenwriter: Interviews With Six Celebrated Screenwriters* (New York: Simon and Schuster, 1981), 279.

[8]Amatasia Toufexis, "Sex Lives and Videotape," *Time,* October 29, 1990, 104.

⁹Nick Ravo, "A Fact of Life: Sex-Video Rentals Gain in Unabashed Popularity," *The New York Times,* May 16, 1990, B1, B8.

¹⁰Dennis Kneale, "Can Fox Cool Down and Stay Hot?" *The Wall Street Journal,* May 5, 1989, B1.

¹¹Montgomery, *Target Prime-Time,* 163.

¹²For example, see CBS' program content standards in Alice M. Henderson and Helaine Doktori, "How the Networks Monitor Program Content," in *Television as a Social Issue,* edited by Stuart Oskamp (Newbury Park, Calif.: Sage, 1988), 130-40.

¹³See, for example, Herbert J. Miles, *Movies and Morals* (Grand Rapids: Zondervan, 1947); Stephen W. Paine, *The Christian and the Movies* (Grand Rapids: Eerdmans, 1957); Kyle Haseldon, *Morality and the Mass Media* (Nashville: Broadman, 1968); Anthony Schillaci, *Movies and Morals* (Notre Dame, Ind.: Fides, 1970). If anyone doubts that Hollywood deserved such moral scrutiny, he or she need only read Kenneth Anger, *Hollywood Babylon* (New York: Dell, 1975).

¹⁴The best of these studies is Rowland, *Politics of TV Violence.*

¹⁵See J. Fred MacDonald, *Blacks and White TV: Afro-Americans in Television since 1948* (Chicago: Nelson-Hall, 1983).

¹⁶Jack Shaheen, *The TV Arabs* (Bowling Green, Ohio: Bowling Green University Press, 1984), 4.

¹⁷Powers, *The Beast, the Eunuch,* 113.

¹⁸Montgomery, *Target Prime-Time,* 217-18.

¹⁹One observer even speculates that pressure groups may ultimately work to the advantage of the networks, program producers and advertisers by reaffirming the industry's essential structure and values. In other words, pressure groups only tinker with a problematic system that is based on liberal capitalism without real social conscience. See Joseph Turow, "Pressure Groups and Television Entertainment: A Framework for Analysis," in Rowland and Watkins, *Interpreting Television,* 142-62.

²⁰"More Sex on TV," *Psychology Today* 22 (1988), 7.

²¹Marianne Paskowski and William Mahoney, "The Making of a TV Uproar," *Electronic Media,* March 6, 1990, 1, 39; "A Mother Is Heard as Sponsors Abandon a TV Hit," *The New York Times,* March 2, 1989, 1, 40.

²²Veith, *Gift of Art,* 39.

²³Myers, *All God's Children,* 95.

²⁴I am indebted here to Nicholas Wolterstorff's fine treatment of this subject in his *Art in Action,* 172-74.

²⁵See Veith, *Gift of Art.*

²⁶Montgomery, *Target Prime-Time.*

²⁷William A. Henry, "When News Becomes Voyeurism," *Time,* March 26, 1984, 64.

[28]For a spectrum of opinion on the movie, and some examples of how it became part of public debate about politics, see Marvin Maurer, "Screening Nuclear War and Vietnam," *Society* 23 (November/December 1985), 68-73; Marvin Kitman, "Blown Away," *The New Leader,* Deember 10, 1984, 22-23; Margaret Ruth Miles, "The Day After," *Christian Century,* March 21-28, 1984, 305-6; Roger L. Shinn, "The Day after *The Day After,"* *The Bulletin of the Atomic Scientists* 40 (February 1984), 43-44; Jonathan Burack, "Kid Stuff," *The Progressive,* January 1984, 50; Marvin Kitman, "In Defense of *The Day After,"* *The New Leader,* November 28, 1983, 21-22; William F. Buckley, "Scoop! *The Day After,"* *National Review,* December 23, 1983, 1632-33.

[29]See William F. Buckley, "The Hitler in Ourselves," *National Review,* March 27, 1987, 62-63; Joseph Vitale, "Made in Amerika," *Channels,* February 1987, 18; Van Gordon Sauter, "A TV Man Views the Storm," *U.S. News and World Report,* February 16, 1987, 68.

[30]Ruth Butler, "Show on Molested Boy Irritates Some Viewers, But Rescues Local Girl," *Grand Rapids Press,* June 18, 1989, H1, H6; Kay Stayner, "I Know My First Name Is Steven," *TV Guide* June 18, 1989, H1, H6.

[31]Dan Nilsen, "Fawcett Is at Her Best as a Battered Wife," *Grand Rapids Press,* October 7, 1984, C1, C2. Also see Neal Karlen, "The Burning Bed," *Newsweek,* October 22, 1984, 38.

[32]I take this phrase from Carl F. H. Henry, *The Uneasy Conscience of Modern Fundamentalism* (Grand Rapids: Eerdmans, 1947).

[33]"Do You Let MTV Answer Your Kid's Most Important Questions?," *U.S. Catholic,* October 1990, 16-21.

[34]C. S. Lewis, *The Screwtape Letters* (New York: Macmillan 1943), 112.

[35]Mark Fackler, "Religious Watchdog Groups and Prime-Time Programming," in Ferré, *Channels of Belief,* 99-116.

Chapter 7: The Soul of Hollywood

[1]Donna Foote, "The Bad and Not So Beautiful," Newsweek, March 25, 1991, 54; Julia Phillips, *You'll Never Eat Lunch in This Town Again* (New York: Random House, 1991).

[2]Benjamin J. Stein, "Hollywood: God Is Nigh," *Newsweek,* December 12, 1988, 8.

[3]Powers, *The Beast, the Eunuch,* 3.

[4]Carnell, *Television,* 19.

[5]Tim LaHaye, *The Battle for the Mind* (Old Tappan, N.J.: Revell, 1980), 152.

[6]Tim LaHaye, *The Hidden Censors* (Old Tappan, N. J.: Revell, 1984).

[7]Linda S. Lichter, S. Robert Lichter and Stanley Rothman, "Hollywood and America: The Odd Couple," *Public Opinion,* December/January 1983, 54-

57. Also see their books *The Media Elite* (Adler and Adler, 1986), and *Watching America: What Television Tells Us about Our Lives* (Prentice-Hall, 1991). Some of their more recent work is summarized in Peter Brimelow, "TV's Killer Businessmen," *Forbes,* December 23, 1991, 38-40.

8James Hitchcock, *What Is Secular Humanism?* (Ann Arbor, Mich.: Servant, 1982), 87.

9Ben Stein, *The View from Sunset Boulevard* (New York: Basic Books, 1979), xii-xiii.

10Quoted in Newcomb and Alley, *Producer's Medium,* 192. Also see Broughton, *Producers on Producing;* and Muriel G. Cantor, *The Hollywood TV Producer: His Work and Audience* (New York: Basic Books, 1971). Since Cantor's book was published there are more female producers, but overall the masculine pronoun in the book's title still applies.

11Quoted in Newcomb and Alley, *Producer's Medium,* 274. Also see June Ammeson, "Garry Marshall Creates Positive Images," *Compass Readings,* April 1990, 71-74.

12Newcomb and Alley, *Producer's Medium,* 219.

13Quoted in Newcomb and Alley, *Producer's Medium,* 72.

14Newcomb and Alley, *Producer's Medium,* 151.

15Richard Levinson and William Link, *Stay Tuned* (New York: St. Martin's, 1981), 68.

16Quoted in Newcomb and Alley, *Producer's Medium,* 83.

17Levinson and Link, *Stay Tuned,* 68.

18Quoted in Newcomb and Alley, *Producers' Medium,* 184.

19Probably the best study so far of the impact a producer can have on the programs he produces is Robert J. Thompson, *Adventures on Prime Time: The Television Programs of Stephen J. Cannell* (New York: Praeger, 1990). Thompson points out that a proven successful producer like Cannell can have moderate control over the networks' decision making.

20David Marc, "Understanding Television," *Harper's Magazine,* August 1984, 34.

21Neil Postman, *Conscientious Objections: Stirring Up Trouble about Language, Technology and Education* (New York: Knopf, 1988), 116-27.

22Muriel G. Cantor, *Prime-Time Television: Content and Control* (Beverly Hills, Calif.: Sage, 1980), 21. Also see Turow, *Playing Doctor,* xvi.

23Scheduling is discussed in many places. See Gitlin, *Inside Prime Time,* 56-62; and Sydney W. Head, *Broadcasting in America,* 4th ed. (Boston: Houghton Mifflin, 1982), 217-20.

24"How Cable Has Changed TV Viewing," *Consumer Reports,* September 19, 1991, 57b.

25"Prime Time's Price Tag," *Channels,* September 10, 1990, 50-51.

26Bob Wisehart, " 'Hill Street Blues' Syndicates for a Fortune," *Grand Rapids*

Press, October 20, 1983, 12B.

[27]*Time,* September 28, 1987. Also see Gary Deeb, " 'The Cosby Show' in Syndication: It's a Powerhouse Already Bringing in Big Bucks," *Grand Rapids Press,* November 21, 1986, B12.

[28]Bob Campbell, "Passion and Intelligence Are Prime Hallmarks of Jodie Foster the Director," *Grand Rapids Press,* November 19, 1991, B4.

Chapter 8: Redeeming the Tube

[1]Readers might like to know that I have been significantly influenced here by the work of William Stringfellow. Although his criticisms of American culture were often blistering attacks on business and especially government, he subtly balanced these diatribes with a profound appreciation for God's grace in the midst of human despair. Some writers dismissed him as a mere "lay theologian." I consider him a prophet in the sense that he discerningly applied biblical perspectives to a critique of contemporary culture and society.

I hope my writing reflects more of God's grace, but I encourage readers to seek out Stringfellow's numerous books. A good one to begin with is *An Ethic for Christians and Other Aliens in a Strange Land* (Waco, Tex.: Word, 1973). Probably the most accessible of his books is *A Simplicity of Faith: My Experience in Mourning* (Nashville: Abingdon, 1982). In the latter he writes that he had "a sense of absurdity—an instinct for paradox—a conviction that truth is never bland but lurks in contradiction—a persuasion that a Hebraic or biblical mentality is more fully and maturely human than the logic of the Greek mind. . . . A biblical person is one who lives within the dialectic of eschatology and ethics, realizing that God's Judgment has as much to do with the humor of the Word as it does with wrath" (86-87). I hope that the same dialectic characterizes my work.

[2]William F. Fore's two books on this subject are among the best critical assessments of the values reflected in much mainstream television: *Mythmakers,* chap. 5; and *Television and Religion,* chap. 4. However, I don't agree completely with Fore's theology of communication. See my review of one of his books: *Calvin Theological Journal* 23 (1988), 100-106.

[3]My wife, a home-health nurse, finds this especially true in the inner city, where video movies are used uncritically as a diversion from street culture and as a mechanism for coping with daily boredom.

[4]One excellent but overlooked book that addresses these issues is Inbody's *Changing Channels.*

[5]Marshall McLuhan and Quentin Fiore, *The Medium Is the Massage* (New York: Bantam, 1967). I do not agree with McLuhan that the fundamental effects of a medium are psychological—that each media technology merely rearranges the priority of the human senses in perception. Nevertheless, the

technology does carry its own bias. By far the best critique of McLuhan's work is James W. Carey, "Harold Adams Innis and Marshall McLuhan," *Antioch Review* (Spring 1967), 5-39.

[6]There are many books and articles describing some of their activities. For an up-to-date list, write to the National Council of Churches, Communication Commission, 475 Riverside Dr., New York, NY, 10115.

[7]Postman, *Amusing Ourselves,* 160-63.

[8]I take this up at greater length in the last chapter of *Televangelism and American Culture.*

[9]See Douglass Cater and Michael J. Nyhan, eds., *The Future of Public Broadcasting* (New York: Praeger, 1976); George H. Gibson, *Public Broadcasting: The Role of the Federal Government, 1912-76* (New York: Praeger, 1977).

[10]Carnell, *Television,* 25.